The ICE Conditions of Contract: Sixth Edition

A User's Guide

Brian Eggleston

CEng, FICE, FIStructE, FCIArb

OXFORD

BLACKWELL SCIENTIFIC PUBLICATIONS

LONDON EDINBURGH BOSTON

MELBOURNE PARIS BERLIN VIENNA

Blackwell Scientific Publications
Editorial Offices:
Osney Mead, Oxford OX2 0EL
25 John Street, London WC1N 2BL
23 Ainslie Place, Edinburgh EH3 6AJ
238 Main Street, Cambridge,
Massachusetts 02142, USA
54 University Street, Carlton
Victoria 3053, Australia

Other Editorial Offices:
Librairie Arnette SA
2, rue Casimir-Delavigne
75006 Paris
France

Blackwell Wissenschafts-Verlag
Meinekestrasse 4
D-1000 Berlin 15
Germany

Blackwell MZV
Feldgasse 13
A-1238 Wien
Austria

First published 1993

Set by DP Photosetting, Aylesbury, Bucks
Printed and bound in Great Britain by
Hartnolls Ltd, Bodmin, Cornwall

DISTRIBUTORS

Marston Book Services Ltd
PO Box 87
Oxford OX2 0DT
(*Orders*: Tel: 0865 791155
Fax: 0865 791927
Telex: 837515)

USA
Blackwell Scientific Publications, Inc.
238 Main Street
Cambridge, MA 02142
(*Orders*: Tel: 800 759-6102
617 876-7000)

Canada
Oxford University Press
70 Wynford Drive
Don Mills
Ontario
(*Orders:* Tel: 416 441-2941)

Australia
Blackwell Scientific Publications
(Australia) Pty Ltd
54 University Street
Carlton, Victoria 3053
(*Orders:* Tel: 03 347-0300)

British Library
Cataloguing in Publication Data
A Catalogue record for this book is available from the British Library

ISBN 0-632-03092-5

Library of Congress
Cataloguing in Publication Data
is available

Contents

Preface

Saying farewell to the Fifth edition is saying farewell to an old friend. It had its faults and so did we but generally we got along together very well. A few worries at the outset about claims; a few problems throughout on interpretation; and a few disputes of note still lingering on from the handful or so of cases which reached the courts over a span of 20 years; but who would deny it was a good contract for reasonable parties and the reasonable engineer? Some of the wording may have been antiquated and some of the phrases obscure and probably few of us reached the end of the 250 word sentence in the war clause but fortunately we never had to face the day war broke out.

Now we must start again and build a new understanding with the Sixth edition. There will of course be suspicions, some expressed in this book, that more has changed than the draftsmen care to admit. There will be disappointment in other quarters that too little has changed and that some much discussed points of contention remain to trouble us for another decade or so. And there will be surprise that the Sixth edition found its way into print with some clauses in fairly obvious need of amendment.

But the Sixth edition starts with a huge advantage over its lately announced rivals. By retaining so much of the form and content of the Fifth edition it has an immediate ease of familiarity and users will not have to go through a learning curve fraught with anxiety and mishaps. This, and its much improved presentation, will ensure its success whatever the debate on the detail.

In writing about the Sixth edition I have had the benefit of many years' use of the authoritative works on the Fifth edition by Duncan Wallace QC and Max Abrahamson and the later writings of John Uff QC in *Keating on Building Contracts*. I have also had the benefit, and enjoyment, of taking part in a score or more seminars on the Sixth edition since its introduction in January 1991 and invariably I have learnt more from the enquiring minds of the delegates than from my own researches. Sadly it is not always possible to give definitive answers to the 'what ifs' and the 'what abouts' and to that problem this book is no exception.

But without the questions we would all be the poorer, so I dedicate this book to all who ask questions and to the three eminent lawyers named above who have done so much to answer them.

Brian Eggleston
5 Park View
Arrow
Alcester
Warwickshire
B49 5PN

October 1992

Chapter 1

Introduction

1.1 *Publication*

The ICE Sixth edition Conditions of Contract were published in January 1991 with the approval of the three sponsoring authorities – the Institution of Civil Engineers (ICE), the Association of Consulting Engineers (ACE) and the Federation of Civil Engineering Contractors (FCEC) – for all works of civil engineering construction.

The Sixth edition will gradually replace the Fifth edition which came out in 1973 and on which it is firmly based. There is however no immediate need for change and the pace of introduction of the Sixth edition is varying, with some employers quickly committed to its use and others adopting a cautious and watchful approach and staying for the time being with the Fifth edition. There is, in any event, a period of overlap which will last for some years as Fifth edition contracts are completed and the slow processes of dispute resolution grind to their fulfilment. But that said, the Sixth edition Conditions of Contract should now be recognised as the principal conditions for use in the civil engineering industry.

1.2 *Competition*

Civil engineers have been remarkably fortunate in having enjoyed for nearly half a century the benefits of using a single dominant form of contract. Just two new editions of the ICE Conditions have been published since 1955 and the sponsoring authorities have shown a commendable degree of discipline in limiting the issue of formal revisions and amendments. The policy has been stability in preference to change and it has paid off handsomely.

There can be little doubt that the singularity and stability of the ICE Conditions of Contract have encouraged civil engineers at all levels to take an interest in contractual matters. And it has been possible to use the contract, safe in the knowledge that what is learnt today will not be redundant tomorrow. In contrast, the building industry is plagued with a multiplicity of standard forms and a constant flood of revisions and amendments: fortunate indeed is the architect, contractor or employer who has two successive jobs under identical conditions of contract. Sadly there are now signs that civil engineering is following the lead of the building industry in developing an appetite for a variety of standard forms. The joint sponsoring authority published the ICE Conditions of Contract for Minor Works in 1988 and an ICE Design and Construct form in late 1992.

Increasing use is being made on civil engineering projects of standard forms published by other bodies although some are clearly intended for other types of work. The Institution of Chemical Engineers Forms of Conditions of Contract for Process Plants (the Red Book and the Green Book) are gaining ground as is the Institution of Mechanical Engineers Model Form of General Conditions of Contract (MF/1) and its Water Services Association variants, the General Conditions of Contract for Water Industry Plant Contracts (Forms P and G/90). The Department of the Environment's General Conditions of Contract for Building and Civil Engineering (GC/Works/1 – edition 3) is also moving into new territory. And then, overshadowing these developments, but promising to lead us back to the path of singularity by virtue of its flexibility, there is the New Engineering Contract produced by the Institution of Civil Engineers on its own initiative. Offered as a radical departure from conventional drafting with flexibility, clarity and the promotion of good project management as its main objectives, it may in the fullness of time prove to be a popular contract.

However there is a long way to go before there is any serious challenge to the ICE Conditions of Contract as the preferred form for the industry. The Conditions are held in such respect and there is such familiarity with their use that it is unlikely they will be displaced within the foreseeable future.

1.3 Background

The First edition of the ICE Conditions of Contract was produced jointly by the Institution of Civil Engineers and the Federation of Civil Engineering Contractors in 1945. Second and subsequent editions have also included the Association of Consulting

Engineers as a sponsoring body. The Second edition was published in 1950; the Third edition in 1951; the Fourth edition in 1955; the Fifth edition in 1973; and the Sixth edition in 1991.

Representatives of the three sponsoring bodies, the ICE, FCEC and ACE, sit on a permanent joint committee – the Conditions of Contract Standing Joint Committee (CCSJC) – which is charged with keeping the use of the ICE Conditions under review and considering any suggestions for amendment. Clearly, from the timespans above, the committee does not embark lightly on the production of a new edition. However the view was taken in the late 1980s that the construction industry had seen great changes since the introduction of the Fifth edition in 1973 and it was appropriate for the committee to undertake a thorough review of the Conditions in the light of new management practices and developments in the industry.

In particular, the committee was drawn to consider:

(a) The implications of the growth of sub-contracting. It was recognised that in the years since the publication of the Fifth edition, large scale direct employment has virtually disappeared from the industry and contractors use sub-contracting as the normal arrangement for constructing the works notwithstanding the restraint in the Fifth edition that sub-contracting is on a consent-only basis.
(b) The debate on the role of the engineer. The position has been under fire from all sides for some years; employers complaining that the engineer is inhibited from adequately representing their interests; contractors complaining that restrictions on the engineer's powers prevent operation of his independence.
(c) The provision of information by the employer. The view has gathered ground in recent years that the discretionary requirement in the Fifth edition for disclosure by the employer of his information on site conditions is not compatible with good practice. Pressure has grown for the employer's obligations to be more positively expressed.
(d) Deficiencies in the clause 14 procedures of the Fifth edition for approval of the contractor's programme. The clause fails to deal with the not uncommon situation of rejection by the engineer.
(e) The introduction of conciliation as a step in the resolution of disputes. The Institution of Civil Engineers published its own Conciliation Procedure in 1988 and introduced its use into the

ICE Minor Works Conditions: the first standard form to move in this direction.

These are but a few of the issues considered by the committee but they are the significant areas of intentional policy change.

1.4 The review policy

It was never intended that the ICE Sixth edition should be anything more than a modernised version of the Fifth edition and the absence of radical change was confirmed in an official release document for the launch of the Sixth edition with these statements:

(a) There has been no change in the principles underlying the Fifth edition.
(b) There has been no significant change in the layout.
(c) There has been no significant change in the allocation of risk between the employer and the contractor.

In approaching the detail of its task, the policy adopted by the drafting committee was outlined by Mr Stuart Mustow, Chairman of the Construction Contracts Standing Joint Committee, who said:

> 'When approaching the review the CCSJC was conscious that clarity and presentation are important and that the Fifth edition left something to be desired in this regard. It was decided therefore to attack turgid prose, bring definitions into one place, break down paragraphs, clarify headings and make the document more attractive and easier to find one's way about. The committee was encouraged in this direction by the standards adopted for the FIDIC Conditions and had already made some progress themselves to this end with the Minor Works Conditions.'

In summarising the actual changes made, the release document listed the opportunities taken in the review as follows:

(a) To shorten and/or break up the text in the interests of readability.
(b) To modernise the language, including some changes in traditional nomenclature.
(c) To take on board some current management practices.

(d) To take on board some recent case law.
(e) To simplify and clarify contractual procedures.

The Sixth edition remains, therefore, a measurement contract based on the engineer's design of the works. There is provision for limited contractor's design but it is not recommended that the Sixth edition should be used as a design and construct form.

Main clause numbers in the Sixth edition correspond to those in the Fifth, although to achieve this the Sixth edition does not use clause 34 which in the Fifth covered the Working Rule Agreement on wages and conditions of operatives. This is now deleted. As a result of breaking-up the text there is an increase in the number of sub-clauses and for these the numbering no longer corresponds exactly. However this is a small price to pay for the improved readability.

1.5 Major changes

The alterations to layout in the Sixth edition may create an illusion of greater change than actually exists and users of the new edition need to exercise care in making comparisons with the Fifth.

A useful publication has been issued by Thomas Telford – *ICE Conditions of Contract 5th and 6th Editions Compared.* This gives the Fifth edition text on left hand pages and the corresponding Sixth edition text on right hand pages with major changes highlighted in pink. The highlighting does not cover all the changes and many significant points are missed or ignored. Nevertheless the publication is of great value in bringing together both editions and readers of this book will undoubtedly find it helpful to have a copy to hand.

The clauses which have had deliberate, significant changes in respect of policy, procedure or drafting are as follows:

(a) clause 2 – the role of the engineer
(b) clause 4 – sub-contracting
(c) clause 8 – contractor's responsibilities
(d) clause 11 – supply of information
(e) clause 14 – contractor's programme
(f) clauses 20–23 – care of the works and insurances
(g) clause 29 – noise, disturbance and pollution
(h) clause 41 – works commencement date
(i) clause 44 – extension of time

(j) clause 46 – accelerated completion
(k) clause 47 – liquidated damages
(l) clause 48 – certificate of substantial completion
(m) clause 59 – nominated sub-contractors
(n) clause 60 – certificates and payments
(o) clause 66 – settlement of disputes

Experience and usage may eventually show other changes to be equally significant but these will have to wait the test of time. All that can be done for now is to identify all changes, whether intentional or not, so that users of the new contract can be alert.

As far as wording is concerned, some overdue changes have been made so that the maintenance period becomes the defects correction period and the forfeiture provisions are renamed determination of the contractor's employment.

1.6 A personal impression

The question of whether or not the Sixth edition is an improvement on the Fifth will be to some extent a matter of personal taste. When the Fifth edition came out in 1973 it was derided in some quarters – of which more is said later in this chapter – but it soon gained widespread use, approval and respect and much of the criticism was seen in retrospect to be misplaced. A similar response greeted the main building form, JCT 80, even though the form it was superseding, JCT 63, was subject to sustained ridicule.

I am conscious therefore that it is easier to criticise than to be constructive; that defects attract and deserve comment whereas good points are all too often left to speak for themselves. Therefore, where there are two views on a particular subject I have tried to express them both. Nevertheless since much of this book is inevitably and quite rightly involved with close examination of the new text more may be said by way of criticism than by praise – but that is detail. By and large I believe the drafting committee achieved most of its objectives even if it has inadvertently left a number of points for us to ponder.

In my view the Sixth edition is not only more readable and easier to follow than the Fifth, but it is better presented, better arranged and it is going to be a better document to use.

However, I do not accept the official claim that there has been no change in the balance of risk between the parties – I have more to say on this in the next section. Nor do I believe that the new wording always achieves the intentions of the draftsmen. There are in fact a few amendments I recommend employers should make before using the contract. However, at the outset all new contracts come under scrutiny and none was ever found to be perfect.

I do have one area of worry about the Sixth edition and that is that the draftsmen, in rearranging the text in the interests of readability, may have significantly altered certain provisions by accident. Consider as an example the contractor's obligation in clause 11 to inspect and examine before submitting his tender. In the Fifth edition the qualification 'so far as is practicable' applied only to the ground and subsoil; in the Sixth edition it applies to the extent and nature of work, the materials necessary for the works, and the means of communication and access to the site. The contractor is then deemed to have based his tender on the inspection and examination so made. Can such a change have been made intentionally with the potential it gives for a new avenue of contractor's claims?

1.7 Balance of risk

For the parties to the contract the balance of risk is likely to be of more importance in the assessment of the merits of a new contract than the niceties of administrative and procedural characteristics.

The official line of the sponsoring bodies is that there has been no significant change in the allocation of risk between the employer and the contractor. In that there has been no deletion or addition of any claim clause this is true, and in that the principal obligations of the parties remain broadly unchanged it is also true. But on a broader interpretation of what is meant by balance of risk one can see that many of the changes introduced in the Sixth edition do either carry additional financial risks for one party or otherwise alter the position of the parties to the benefit of one and the detriment of the other. A simple example of the first is that, under clause 29(4), the employer must now indemnify the contractor against noise, disturbance and pollution claims which are the unavoidable consequence of carrying out the works. An example of the second is the removal from clause 47 of the conditions precedent to deduction of liquidated damages by the employer.

In an appendix at the end of this book I have scheduled all the significant changes between the Fifth and Sixth editions in what I term an 'improvement analysis'. Some are financially related, some are time related and others are matters of administrative convenience. In a simple number count I find 67 improvements in the contractor's position against 31 in the employer's position. Some are indeed small but when added together they cannot be dismissed as of no significance.

Even taking into account the argument that a number count is inconclusive in itself I think it difficult to avoid the impression that the contractor has done better out of the changes in the Sixth edition than the employer. This is a long way from the charge once levelled at a JCT form that an architect who advised his client to use it might be guilty of professional negligence. It is also a long way from the charges made on the publication of the ICE Fifth edition of contractor bias. But the parties are entitled to know how their position has changed on the introduction of a new form so they can make their decision on its use, with or without amendment, on a reasoned basis.

1.8 A look back to the introduction of the Fifth edition

In 1969 a joint committee from the sponsoring bodies was appointed to undertake within a period of 12 months 'a modest revision' of the ICE Fourth edition Conditions of Contract which had then been in use for fourteen years. Four years later, in 1973, after 66 meetings of the committee and 70 meetings of the working party, the 'modest revision' was born as the ICE Fifth edition.

The initial response, particularly from the legal profession, to the Fifth edition was unashamedly hostile and its introduction was greeted with a rash of articles in the technical press under such headlines as:

> 'Bad drafting sires a lawyer's gift horse'
> 'The model revision which became a torrent of change'
> 'A contractor's charter to riches'

So fierce was the criticism, and so distinguished were the critics, that Sir William Harris, Chairman of the Joint Contracts Committee and David Gardam QC, legal adviser to the committee, were stung into publishing an eight page reply in *New Civil Engineer*, 20 December 1973, under the title 'Clearing the critics' confusion'.

Looking back now after 19 years in which the Fifth edition has been accepted on all sides as an unqualified success, it is difficult to see what all the fuss was about.

Certainly there was a fear that clause 13(3) (engineer's instructions) and clause 56(2) (increase or decrease of rate) would open the door to a flood of claims but there is no evidence that either has been unduly troublesome even if the potential exists. No doubt the lawyers were right with many of their comments on flaws in drafting but perhaps they under-estimated the capacity of civil engineers to cope with such matters.

The best explanation for the fuss was probably that advanced at the time by Sir William Harris and David Gardam QC in their article in *New Civil Engineer*:

> 'it appears to be a fundamental disagreement [on principles] which lies at the root of much of the criticism.'

The leading critic of the new form, I. N. Duncan Wallace QC, had suggested, in full consistency with his writings elsewhere, in an article in *New Civil Engineer* on 1 November 1973 that:

> 'The overriding commercial interest of an employer is to know, as accurately as possible, barring variations ordered by his advisers, fluctuations clauses, and any quite exceptional risk which he is prepared himself to shoulder, the full extent of his financial commitment before the start of a project. If the contract enables him to know this, he can either avoid committing himself at all, or can effect any necessary savings beforehand.
>
> This commercial need is far outweighed by any consideration of a keener tender price let alone a price which is only a price in name. It is as much in the interests of the contracting side of the industry as of employers that this basic reality should be understood.'

This was, and is, very true, no doubt, for a lump sum building contract, but how applicable was that proposition to a measurement type civil engineering contract? And here it should be added, one major aspect of the Fifth edition was that it was clearly revealed as a measurement contract and not a lump sum contract as many believed the Fourth edition to be.

The answer to this is again best left to the article by Sir William Harris and David Gardam QC with this exposition of the policy of the ICE Fifth edition and its principles:

> 'It is a function of a contract to define upon whom the various risks of an enterprise shall fall, and it was decided that the Contractor should only price for those risks which an experienced contractor could reasonably be expected to foresee at the time of tender....
>
> It is the right and duty of the Employer to decide and, by his Engineer, to design and specify that which is to be done and it is the Employer's duty to allow the Contractor to do that which is to be done without hindrance.
>
> It is the duty of the Contractor to do what the contract requires to be done, as designed and specified by the Engineer, but, subject to any specific requirement in the contract, it is his right and duty to decide the manner in which he will do it.
>
> If there are to be exceptional cases where the Contractor is to decide what to do, or to design what is to be done, or where the Employer or the Engineer is to decide how the work is to be done, the contract must expressly provide for this and for the necessary financial consequences for the protection of the Contractor.'

This exposition remains as sound for the Sixth edition as for the Fifth and it remains as good a starting point for the analysis of any contractual problem as to be found anywhere.

1.9 *Reception of the Sixth edition*

The reception of the Sixth edition has been decidedly muted compared with that accorded to the Fifth edition. Just an occasional article here and there pointing out a few quirks and areas of difficulty but not enough debate to generate heated controversy.

This may be in part because the improvements in clarity and presentation are so obvious and it may be because this time there are no fundamental misunderstandings on policy and principles.

At seminars on the Sixth edition there have been two subjects which attract most interest:

(a) the revisions to clause 4 and the freedom of the contractor to sub-contract without approval of the engineer, and
(b) the revisions to clause 11 and the obligations on the employer for disclosure of information.

Both are very much rooted in practical considerations close to the hearts of engineers and for the time being the lawyers are holding their fire. Perhaps some of the comments in this book will arouse their interest.

1.10 This user's guide

In writing this user's guide, which is intended very much as a practical work, I have assumed that the reader has a basic knowledge of contract law, some familiarity with the Fifth edition and, most importantly, since I have avoided quoting the clauses of the Conditions in full, a copy of the Sixth edition to hand. The absence of capital letters from the contractor, employer, engineer and other titles is a matter of style to assist readability although not everyone will approve of such liberty taking. Generally I have taken the clauses in numerical order and in chapters corresponding to the sections of the Conditions but a few exceptions to this approach have been necessary.

For each clause I have illustrated differences from the Fifth edition and offered a view, or views, of purpose, interpretation and problems of usage. Where I know there are conflicting views on matters of some importance I have endeavoured to put both sides of the argument and occasionally I have been drawn to the unavoidable conclusion that no amount of analysis can produce a definitive answer. More than once I have been forced to offer the advice: do not move until you have consulted your lawyer. Disappointing, perhaps, that this should be so but we live by a system where the law is uncertain until it has spoken and then we are left with its decision for good or for bad unless and until it is overturned.

For those contractual provisions which have had the benefit of firm legal rulings or relevant judicial guidance, I have outlined the law and the cases of interest. Whenever possible I have taken an extract from an appropriate judgment to illustrate a point of importance because I believe it assists understanding of the Sixth edition Conditions to see how judges arrive at their decisions, albeit sometimes on issues under other conditions.

Finally my advice to all users of these and other conditions is never to act or make a decision on any contractual matter without first reading and re-reading the relevant clauses and then seeing what two or more reference works have to say on the subject. Very few of us when we use conditions of contract are dealing with our own money and we usually owe it to someone to do our best for them. I hope that this book helps you.

Chapter 2

Definitions and notices

2.1 *Introduction*

This chapter covers clause 1 (definitions and interpretation) and clause 68 (serving of notices).

Clause 1 gives, with varying degrees of clarity and helpfulness, definitions of 24 words and phrases used in the Conditions. It also adds for the purposes of interpretation that words importing the singular also include the plural and vice versa and that headings and marginal notes are not to be taken into consideration in the interpretation or construction of the contract. A new sub-clause 1(6) recognises advances in information technology and permits communications in writing to be sent by hand, post, telex, cable or facsimile.

The list of 24 definitions shows an increase of eight over the Fifth edition but this is largely a matter of presentation since seven of the additional definitions are to be found scattered in the main text of the Fifth edition. The one new definition is that the 'Bills of Quantities means the priced and completed Bills of Quantities' – a point previously covered in the definition of the 'Contract'.

2.2 *Identification definitions*

Employer and engineer

The names of both the employer and the engineer are now to be stated in the appendix to the form of tender. This removes the scope for an administrative slip which could easily occur under the Fifth edition where the parties names were supposed to be inserted in the space provided in clause 1 of the Conditions but frequently were not.

Both the employer and contractor definitions include, as in the Fifth edition, personal representatives and successors. This clarifies the point that the contract continues in force in the event of death of a party who is a person; or legal change of ownership of a firm or company; or formal change of a statutory body by reorganisation or amalgamation.

As to the position where a firm goes into administration or administrative receivership see clause 63 (determination). The employer may determine the contractor's employment but this does not discharge the contract.

Employer

The phrase 'permitted assigns' has been added to the definition of the employer. In the Fifth edition the phrase only applied to the contractor. This supports a change in clause 3 (assignment) which in the Sixth edition permits assignment with consent by either the employer or the contractor.

Contractor

A minor but worthwhile change of wording has been made; 'Whose tender has been accepted' becomes 'to whom the Contract has been awarded'.

The form of tender of both editions states that written acceptance of 'this Tender' constitutes a binding contract. However there can be problems in proceeding on this basis. The standard form of tender may not have been used or the acceptance may have been qualified. In either case a contract may not have been formed merely by the process of submission of a tender and despatch of a letter of acceptance – see the comment in section 6.3.

Engineer

The definition of engineer now expressly covers 'the person, firm or company appointed by the Employer'.

It has, of course, long been the practice for firms to be appointed as engineers under earlier editions of the Conditions without such wording. The practice, though, has been open to criticism not least on the grounds that some of the clauses imply an individual approach to judgement. However the new requirement in clause 2 for the engineer to be a named individual resolves this point, so that any objection to the engineer being a firm is now effectively removed.

The Sixth edition retains the employer's power to reappoint – an essential power without which the employer might be in some difficulty if the appointed engineer were to die or retire (if a person) or go out of business (if a firm). However, the wording also gives the employer the right to replace his appointed engineer – a right which is granted to the employer in some standard forms of contract only with the contractor's consent. The employer is not obliged to give reasons for reappointing.

The power of the employer to reappoint also creates a duty as an implied term if the appointed engineer is no longer able to continue. In *Croudace Ltd* v. *London Borough of Lambeth* (1986) it was held that Lambeth was in breach of contract in failing to nominate a successor to an architect who had retired, thereby leaving no one legally able to issue certificates.

Engineer's representative

The definition of the engineer's representative in the Fifth edition was unnecessarily detailed in its reference to 'resident engineer', 'assistant' or 'clerk of works' and the clause was capable of a variety of interpretations.

The definition in the Sixth edition sensibly refers only to a person notified by the engineer under clause 2(3) (a).

2.3 *Contract document definitions*

As with the Fifth edition the contract is defined to include:

- Conditions of contract
- Specification
- Drawings
- Bill of quantities
- Tender
- Written acceptance of tender
- Contract agreement (if completed)

No priority is given to any particular document for purposes of interpretation unlike JCT forms where nothing in the bills of quantities is permitted to override or modify the agreement, conditions or appendix. Instead, in both the Fifth and Sixth editions, the documents are taken to be mutually explanatory and the engineer is required under clause 5 to explain and adjust any ambiguities or discrepancies.

Conditions of contract

It is quite usual and fully permissible in civil engineering contracts for the conditions of contract to be incorporated by reference only. In the form of agreement they are deemed to form part of the contract and in the form of tender they are similarly included. However, in the building industry the practice is to include a copy of the conditions in the bundle of contract documents and for each and every amendment to be signed or initialled by both parties. If amendments are made to the standard ICE Conditions it is wise to follow the same procedure.

Changes to the conditions of contract, once the contract has been let, are not usually envisaged. Clearly the parties have the power to make such changes by agreement and the new provision for accelerated completion in clause 46(3) of the Sixth edition is a rare example of provision for such change being made in the contract itself. However, neither party can unilaterally change the terms and conditions nor can the engineer make such changes – a matter previously implied but now expressly stated in the new clause 2(1) (c).

Specification and drawings

The definitions of specification and drawings include documents referred to in the tender and modifications or additions issued or approved in writing by the engineer.

There are three separate issues to consider here:

(a) Specifications and drawings which are necessary to enable the contractor to complete the works but which do not give rise to any variation. The engineer has power to deal with these under clause 7(1) – (further drawings, specifications and instructions). The issue of these is not in any way incompatible with the finality of the conditions of contract referred to above. They do not vary the contract or even the contract works. They supply necessary detail.

(b) Specifications and drawings which amount to variations. The engineer has power to deal with these under clauses 7(1) and 51(1). Again these do not vary the contract – only the works.
(c) Specifications or drawings approved in writing by the engineer. This is a dangerous phrase but it is no different in the Sixth edition than in the Fifth. It has the effect of giving contractual status to any specification or drawing approved in writing by the engineer, whether supplied by the engineer or the contractor. The definition is not restricted to the permanent works. Clearly the engineer must exercise great care in approving any specifications, drawings or proposals for change put forward by the contractor. More is said on this in later chapters.

2.4 *Financial Definitions*

Bill of quantities

The bill of quantities has been defined in the Sixth edition as meaning the priced and completed bill of quantities. It avoids the drafting problems in the Fifth edition where the contract definition referred to the 'Priced Bill of Quantities' but elsewhere the payment and measurement clauses referred to the 'Bill of Quantities'.

Tender total

In measurement contracts such as the ICE Fifth and Sixth editions, the tender total has no obvious contractual status. It may be helpful to the employer in comparing tenders but it is not effective in determining the contract price.

However in both editions there is reference in the appendix to the tender total in connection with limits of retention and there is a further reference in the appendix and in clause 10 in relation to performance bonds. For these matters, at least, the definition is useful.

The Sixth edition extends the wording of the definition of tender total by providing, as an alternative to the total of the bill of quantities and in the absence of such a bill, 'the agreed estimated total value of the Works'.

This presumably relates to a schedule of rates for otherwise it is not clear how the Sixth edition, or for that matter the Fifth edition, could work without a bill of quantities. It is clearly not drafted as a lump sum contract and it would be difficult to operate on an activity schedule basis.

Contract price

There is no change in the definition of contract price between the Fifth and Sixth editions. The contract price remains the sum to be ascertained and paid in accordance with the provisions.

Since more than half the 71 clauses in the contract have financial implications the contract price is likely to be indeterminate until the works are completed. The phrase, therefore, is rarely used in the Conditions and its main purpose is to fix in the form of agreement the consideration due to the contractor.

Payments made to the contractor for breaches of contract by the employer not provided for in the Conditions do not form part of the contract price. An interesting point to consider when it comes to calculation of retention money.

2.5 *Prime cost and provisional sums*

In the Fifth edition definitions of prime cost items and provisional sums were to be found in clause 58. In the Sixth edition they are grouped with other definitions in clause 1.

Prime cost

There is no change of the wording used in the Fifth edition. The definition itself is circular and only acquires meaning in the context of clause 58(2) – (use of prime cost items). This is discussed in chapter 16.

Provisional sum

This definition has been redrafted to make it clear that a provisional sum relates only to a specific contingency. It was possible to read the definition in the Fifth edition as including general contingency sums.

However there is no change in the general principle that a provisional sum may be used in whole or in part or not at all at the direction and discretion of the engineer.

2.6 *Nominated sub-contractor*

This is another definition brought forward from clause 58 of the Fifth edition to clause 1 of the Sixth.

It remains essentially unchanged, but in keeping with the policy in the Sixth edition of avoiding cross references wherever possible, it is redrafted to state directly that a nominated sub-contractor's employment arises only under a prime cost item or a provisional sum.

The definition makes it clear that, if a nomination of a supplier or sub-contractor is made other than through a prime cost item or provisional sum, it is not a nomination covered by the provisions of clause 59. So if there is a named supplier in a specification item or variation this is not to be regarded as a nomination.

Named suppliers

There are potential difficulties for both employer and contractor in using named suppliers or sub-contractors outside the scope of clause 59 since the practice cuts across the contractor's ordinary responsibility. Thus if a named supplier goes out of business or refuses to trade with the contractor, the engineer is obliged to give instructions or variations at the employer's cost on how the works should be completed.

As to sub-contractors it is sometimes argued that the contractor has a right to do all work other than that which is contractually nominated and that naming sub-contractors outside the scope of clause 59 is not permitted. However, it has to be recognised that the engineer's powers of instruction and variation under clauses 13 and 51 are very strong and wide and the contractor is contractually obliged to comply with whatever the engineer requires.

Faults in named products

The contractual position when a named product is unsatisfactory is complex in any contract, and under both the Fifth and Sixth editions it may depend on how the naming is made. Thus instructions given under clause 13 give the contractor entitlement to recovery of costs beyond those which could have been foreseen at the time of tender. This would seem to be all the costs arising from use of an unsatisfactory product. However, for products named in the tender documents or by variation the contractual position may be different.

Two decisions of the House of Lords both given on the same day are instructive. In *Young & Marten Ltd* v. *McManus Childs Ltd* (1969) it was held that a sub-contractor was responsible for latent defects in a batch of faulty 'Somerset 13' roofing tiles. It was said that naming the tiles excluded the warranty that they would be fit for their purpose but did not exclude the warranty that they would be of merchantable quality.

In *Gloucestershire County Council* v. *Richardson* (1969) it was held that the contractor was not responsible for latent defects in precast concrete columns supplied by a named manufacturer. The distinction between the two cases was made on the grounds that in the first the sub-contractor, although restricted to a brand article, was free to agree his terms of trade, whereas in the second case the contractor had been instructed to accept a quotation on the manufacturer's terms and had no rights of objection as he would have had for a nominated sub-contractor on the grounds that the manufacturer would not provide an indemnity.

The lesson from these cases is that the contractor should try to obtain from any named suppliers or sub-contractors the same indemnities he would seek if they were contractually nominated, and then if such indemnities are not available he should ask the engineer for instructions on how to proceed.

Options and alternatives

To avoid the difficulties described above specifications sometimes name suppliers either with an option or with the phrase 'or other approved'. The two expressions give different legal rights.

In *Leedsford Ltd* v. *Bradford City Council* (1956) it was held that the conduct of an architect in rejecting alternatives put forward by the contractor under an item for 'Empire Stone Company or other approved' was not breach of contract by the employer. There was an absolute obligation on the contractor to supply Empire Stone unless the employer, through the architect, approved some other stone. The employer was neither obliged to approve some other nor give reasons for rejection.

In *Crosby & Sons Ltd* v. *Portland Urban District Council* (1967) however, it was held that the contractor was entitled to a variation when the engineer instructed that pipes should be 'Staveley' thereby depriving the contractor of the choice in the bill item which stated that pipes were to be 'Stanton or Staveley'.

2.7 *The works*

The definitions of 'Permanent Works', 'Temporary Works' and 'Works' remain essentially the same in the Sixth as in the Fifth edition. They have been much criticised for their circularity and lack of precision. It is clear that some items such as cofferdams or grouting might fall into both the permanent and temporary categories.

The question of whether or not the distinction matters revolves around the use of the defined phrases in the Conditions. Generally the phrase used is 'the Works' which by definition includes both permanent and temporary works. However a few clauses do rely on a distinction, notably:

- clause 8(2) – design responsibility
- clause 14(5) and (6) – methods of construction
- clause 20 – care of the works
- clause 53 – vesting of equipment
- clause 60 – payments

The intention of the distinction is usually clear although some observations are made in chapter 7 in relation to clause 20 where there are references to 'Permanent Works' which seem oddly restrictive.

A stronger criticism might be that the phrase 'the Works' is used in certain clauses, particularly clauses 13 and 51 (instructions and variations), which could place financial obligations on the employer in connection with deficiencies or changes in temporary works. This is discussed further in chapter 6.

Works commencement date

This is a new definition arising from an improvement in the

drafting of clause 41. The detail of the definition remains in that clause.

2.8 *Certificates*

The Fifth edition had within its text references to the 'Certificate of Completion', the 'Period of Maintenance' and the 'Maintenance Certificate'.

These phrases have been retitled and brought forward to clause 1 as definitions.

Certificate of substantial completion

This is defined in clause 1(1) (r) as a certificate issued under clause 48. The insertion of the word 'substantial' is welcome in correcting an oddity in the Fifth edition where the provisions for substantial completion led to the issue of a 'certificate of completion'. The legal distinction between completion and substantial completion is discussed in chapter 12.

Defects correction certificate

The inappropriate use of the word 'maintenance' has been dropped from the Sixth edition. The 'Maintenance Period' has become the 'Defects Correction Period' and the 'Maintenance Certificate' has become the 'Defects Correction Certificate'.

2.9 *Sections*

The definition of a 'Section' in clause 1(1) (u) of the Sixth edition is the same as that in the Fifth. Both refer to parts of the works separately identified in the appendix to the form of tender.

This gives a clear distinction between 'sections' of the works and 'parts' of the works. The contractor is under no obligation to finish early any part of the works which is not identified as a section in the appendix although the Conditions do provide that a part can attract its own certificates and defects correction period.

2.10 *The site*

The site was another definition much criticised for its vagueness in the Fifth edition and it is not much improved in clause 1(1)(v) of the Sixth edition.

The question of whether the phrase 'provided by the employer' used in both editions applies to 'lands and other places on, under, in or through which the Works are to be executed' as well as to 'any other land or places' is not settled. If it does not, as has been suggested by some commentators, the site could arguably include areas and premises remote from the location of the permanent works and well outside what would normally be understood as 'the Site'. However it is difficult to reconcile this interpretation with the obligation in clause 42 for the employer to give possession of 'the Site'. That clause certainly implies, in both the Fifth and Sixth editions, that the employer has control over the site.

Fortunately the word 'site' does not appear often in the Conditions and the scope for contention is limited. Payment for materials under clause 60 is the most obvious point of conflict.

There is, in any event, an important change in the definition of the site in the Sixth edition with the addition of the words 'together with such other places as may be designated in the Contract or subsequently agreed by the Engineer as forming part of the Site'.

This allows the site to be extended to land beyond that provided by the employer to include such areas as the contractor's compound, borrow pits, production facilities and even access routes. There is no obvious advantage to the employer in this but the contractor could benefit by easier qualification for payment under clause 60 for materials 'delivered to the Site'. The engineer would need to consider the financial risk implications of this before agreeing to extending the site.

2.11 *Contractor's equipment*

The definition in clause 1(1)(w) replaces with almost identical wording the definition of 'Constructional Plant' in the Fifth edition. The change follows through into the main body of the Conditions in clause 53 and elsewhere. It removes the source of confusion between plant which is part of the permanent work – pipework, mechanical and electrical equipment etc. – and contractor's plant in the sense of excavators or scaffolding.

2.12 *Cost*

The application of the definition of 'Cost' given in the Fifth edition led to much argument and this is not likely to be reduced by the extended definition now given in clause 1(5) of the Sixth edition.

Overhead cost in the Fifth edition

The difficulty in the Fifth edition revolved around the phrase 'cost ... shall be deemed to include overhead costs whether on or off the Site'. Did this mean that the contractor could automatically add a percentage for overhead cost to his prime cost or did it mean that overhead costs could be included where incurred and provable? Contractors quite naturally went for the first interpretation and sceptical engineers for the second.

On ordinary legal principles the contractor should not be entitled to recover any element of cost that is not incurred (or otherwise suffered within the meaning of loss and expense). On this basis the contractor would have to prove his cost. However contractual provisions do not have to follow ordinary legal principles and it may have been intended under the Fifth edition that the contractor should have an automatic right to an overhead percentage.

In favour of this is the phrase 'cost when used in the Conditions of Contract' which suggests that for common law claims (or extra-contractual claims as they are sometimes called) a different definition of cost might apply. Against the automatic right argument however is the practical point that even if off-site overheads are allowed without scrutiny the same would rarely apply to on-site overheads, so why make a distinction?

Cost in the Sixth edition

The Sixth edition seems to have resolved the dispute on whether cost need be incurred – indeed that is how it is now defined – but is still not clear if 'cost' is the same as 'loss and expense' as defined in building contracts and interpreted in the courts.

'Expenditure properly incurred'

This phrase in clause 1(5) seems to put beyond doubt that the contractor can only recover the cost he has incurred. However is the later phrase in the clause 'and other charges properly allocateable thereto' caught by this or does it provide an opportunity for the contractor to allocate charges which are not strictly expenditure arising from the claim but are, for example, on-going items such as head office rates?

If the correct interpretation is that all expenditure has to be incurred, does this prevent the contractor from recovering lost overheads when prolongation occurs (see later in this chapter for the difficulties of calculating such overheads)? If it does, the contractor's position is worse than it would be at common law for the recovery of loss and expense. Perhaps this is the intention of the contract but it raises the difficult question, examined further in chapter 13, as to whether the express words of the Conditions are sufficient to exclude the contractor's common law rights or whether these remain as an alternative remedy.

'Or to be incurred'

This is a curious phrase and it is not clear why it has been included in clause 1(5). To see if it has any effect it is necessary to look at the way the word 'cost' is used in the Conditions.

Clause 7(4) (a), for example, says the contractor is to be paid, in accordance with clause 60, the amount of such cost as may be reasonable. Clauses 13(3), 31(2) and 42(3) say likewise. Clauses 12(6), 14(8) and 40(1), however, say the contractor shall be paid the cost incurred. Clause 60, the payment clause, says the contractor shall submit at monthly intervals a statement showing the amounts he is entitled to under the contract.

It is certainly odd if the contractor is required or is allowed to claim under some clauses for cost which is still to be incurred, whereas under others he is to claim only for costs which have been incurred. It is also a matter of concern to both the engineer and employer if the contractor is entitled to payment for costs still to be incurred. The comment in chapter 18 on interest on overdue payments shows how difficult the engineer's task is in correctly certifying amounts due on claims and how the employer's liability for interest can arise unexpectedly. If there is a weakness here in the Conditions, then contractors may not be slow to exploit it.

Remoteness of cost

One question on the phrase 'cost ... means all expenditure properly incurred', as used in clause 1(5), is whether it can be taken to avoid the rules for remoteness of damage developed from the case of *Hadley* v. *Baxendale* (1854). In other words, if the contractor can show that he has incurred a cost, however remote, can that cost be recovered under clause 1(5) even though it would not be allowed in a claim for damages? The answer may lie in the word 'properly' which clearly has some restrictive effect.

'Overhead ... does not include any allowance for profit'

The Fifth edition did not expressly state that cost did not include profit.

In general, where cost is recoverable for breach of contract or under a similar contractual provision, profit is not normally included. However for additional work (rate comparability aside) and for provisions such as clause 12 for unforeseen conditions, the inclusion of profit is compatible with general principles. The Sixth edition allows for this by including profit on extra permanent or temporary work in each of its claim clauses.

The Sixth edition, therefore, deals in a more logical way with profit than did the Fifth. There remains, however, a problem with overheads on whether or not it is possible to isolate profit.

On-site overheads should usually be straightforward to calculate although it has to be recognised that contractors vary widely in their treatment of overheads. Large contractors tend to apportion as many costs as possible to each contract whereas small contractors not infrequently make no apportionment at all – not even for plant or supervision. Consequently since there is no rule for what constitutes on-site, as opposed to off-site, overheads each analysis has to be approached with at least one eye on the possibility of overlap.

The difficult problem is with off-site overheads. There are many aspects to this. Firstly only rarely will the contractor's head office overheads actually increase as a result of a claims situation on a particular contract. If extra cost is the basis for recovery there will rarely, therefore, be anything to recover.

However, if allocation is a permissible basis of recovery, as it would seem to be on general principles from the words of Mr Justice Forbes in *Tate & Lyle Foods Distribution Company Ltd* v.

Greater London Council (1982) the contractor has an entitlement. He said:

> 'I have no doubt that the expenditure of managerial time in remedying an actual wrong done to a trading concern can properly form the subject matter of a claim.'

However, the judge went on to reject the 2.5% claimed for managerial cost and said it was up to managers to keep time records of their activities.

Lost overheads

A further problem relates to claims for loss of contribution to overheads – a situation which is commonly created by delay in completion of the works. Such claims are acceptable at common law as confirmed by Sir William Stabb QC in *J. F. Finnegan Ltd* v. *Sheffield City Council* (1988). He said:

> 'It is generally accepted that, in principle, a contractor who is delayed in completing a contract, due to the default of his employer, may properly have a claim for head office or off-site overheads during the period of delay, on the basis that the workforce, but for the delay, might have had the opportunity of being employed on another contract which would have had the effect of funding overheads during the overrun period.'

However the calculation of the amount due is not only difficult in itself, although various formulae have been devised and accepted by the courts in appropriate circumstances, but there is frequently no accurate way of separating out what might be termed profit. In the few reported cases the courts do not appear to have attempted to do so.

The point is that while it is a simple accountancy exercise to calculate the gross profit on a contract (value less cost), the calculation of net profit can only be a theoretical exercise. Net profit for the business is the balance between combined gross profits and overheads but net profit for a contract depends on allocation of overheads. This in turn depends on the turnover from other contracts. Consequently what could be a net profit on a contract for one contractor could be a net loss for another, although the circumstances of the contracts and the levels of overheads of both contractors could be identical.

Whichever way the approach to off-site overheads is made under the Sixth edition, problems are encountered. Neither contractors nor employers will benefit from this.

Financing charges

Interest and financing charges may seem on the face of it to be the same thing but in law they are different.

Interest charges relate to late payment of a sum due; and generally at common law it is well settled that debts do not carry interest.

Financing charges relate to costs (or loss and expense) which form the subject of claims for the period from which such costs are incurred to the time of application for payment or certification.

The principle concerned in the payment of financing charges is that the contractor has incurred expense in financing the primary expense involved in his claim and this financing charge, therefore, is not interest on a debt but a constituent part of the debt.

The principle was established by the Court of Appeal in the case of *F. C. Minter Ltd* v. *Welsh Health Technical Services Organisation* (1980). Lord Justice Stephenson put the matter this way:

> 'It is further agreed that in the building and construction industry the "cash flow" is vital to the contractor and delay in paying him for the work he does naturally results in the ordinary course of things in his being short of working capital, having to borrow capital to pay wages and hire charges and locking up in plant, labour and materials capital which he would have invested elsewhere. The loss of interest which he has to pay on the capital he is forced to borrow and on the capital which he is not free to invest would be recoverable for the employer's breach of contract within the first rule in *Hadley* v. *Baxendale* (1854) 9 Ex 341 without resorting to the second, and would accordingly be a direct loss, if an authorised variation of the works, or the regular progress of the works having been materially affected by an event specified in clause 24(1) has involved the contractor in that loss.'

The decision in *Minter* was followed by the Court of Appeal in *Rees and Kirby Ltd* v. *Swansea City Council* (1985) where it was confirmed that calculation of financing charges should be on a compound interest basis.

It is now generally accepted, following *Minter* and *Rees and Kirby* that, under most standard forms of construction contracts, the contractor is entitled to include in his application for loss and expense or extra cost the financing charges he has incurred up to that time. The Fifth edition remained silent on the point to the end although there was little serious opposition to payment by the late 1980s.

The Sixth edition expressly includes finance charges as an element of cost in clause 1(5) so arguments on principle have been eliminated. There is however still scope for argument on the matter of whether it is necessary for the financing charges to be 'incurred'. No distinction was made in *Minter* between the cost of borrowed money and the cost of the contractor's own money but some will say that only with borrowed money can the cost be said to be 'incurred'.

2.13 Clause 68 – notices

Clause 68 remains as difficult a clause to interpret in the Sixth edition as it was in the Fifth. It deals with notices to be given to the contractor and notices to the employer.

Clause 68(1) provides that all notices to be given to the contractor shall be served in writing at the contractor's principal place of business, and for a company at its registered office. Whilst it is understandable that formal notices such as those under clause 63 (determination) and clause 66 (engineer's decisions) should be so treated it is hard to accept, for example, that when the engineer requires the contractor to attend for measurement under clause 56(3) he must write to the contractor's registered office. The problem arises because under clause 15(2) the contractor's agent is only empowered to receive 'directions and instructions' from the engineer whereas many of the practical clauses of the Conditions require the engineer to give 'notice' to the contractor. For the avoidance of doubt it is suggested that this clause requires amendment to distinguish between serving formal notice on contractual matters and giving practical notice on constructional matters.

Clause 68(2) covers notices to be given to the employer and these generally are few and formal. The clause does not apply to notices given to the engineer by the contractor.

Chapter 3

The engineer

3.1 *Introduction*

This chapter examines clause 2 of the Conditions which deals with the duties and authority of the engineer and the function and powers of his representative and assistants.

The Sixth edition continues the policy of the Fifth in assigning to the engineer the roles of:

- designer
- supervisor
- contract administrator
- certifier

This requires the engineer to exercise a difficult balance between fulfilling his obligations to the employer when acting as his agent, and carrying out those duties under the contract which demand impartiality.

The Fifth edition was silent on this matter and clause 2 of that edition did little more than set out limitations on the delegation of the engineer's powers and establish the functions of the engineer's representative and any assistants. In the Sixth edition clause 2 is expanded with provisions on the conduct of the engineer and the need for him to be a named individual.

The engineer, however, is not a party to the contract and the consequences of any infringement by the engineer, his representative, or assistants of the provisions of the contract have to be settled directly between the employer and the contractor.

Duty of certifiers

The engineer will, of course, have his own contract of engagement with the employer and under that he can be sued, but only by the employer, for any breach. It was held in the case of *Sutcliffe* v. *Thackrah* (1974) that an employer could recover loss from his architect who had negligently over-certified prior to the contractor going into liquidation. The architect's defence that he was acting in an arbital capacity when certifying was not accepted.

Many legal commentators felt that a similar remedy would be open to a contractor as an action in negligence when the engineer had caused loss by under-certifying. But the Court of Appeal in *Pacific Associates* v. *Baxter* (1988) rejected such a claim holding that the engineer had no duty of care in respect of economic loss to the contractor and that the arbitration clause in the contract provided the contractor's remedy. The decision has been criticised because there had been support in earlier cases, notably *Arenson* v. *Arenson* (1977) and *Sallis* v. *Calil* (1987), for the proposition that a certifier owed a duty of care to the contractor and it will come as no surprise if contractors continue to argue that, if certifiers owe duties to employers, they also owe them to contractors.

3.2 *Duties and authority of engineer*

Obligation to carry out duties

Clause 2(1)(a) of the Sixth edition expressly states that the engineer shall carry out the duties specified in or necessarily to be implied from the contract. The Fifth edition had no corresponding provision although there was little doubt that there was an implied term in the contract to that effect.

Even if the new sub-clause does nothing more than restate the accepted legal position it is still a significant item of drafting change. It can be taken as signalling three messages:

(a) to engineers – as a direction on conduct
(b) to employers – as a warning on liability
(c) to contractors – as to rights of claim.

Its importance lies in relation to those clauses of the Conditions where the engineer has duties to perform but there is no provision in those clauses for the contractor to claim costs for breach.

Consider, as an example, the examination of work before covering-up. Under clause 38(1) of both the Fifth and Sixth editions, no work is to be covered up or put out of view without the consent of the engineer; the contractor is required to give notice; and the engineer is to attend without delay unless he advises the contractor he considers it unnecessary. Theoretically, at least, the contractor is barred from progressing until the engineer has carried out his duties under the clause. In practice things often work out differently with the contractor carrying on regardless, fearful that he can find no clause in the contract to support a claim for delay.

Under the Fifth edition such a claim would have to be based on an implied term – and that would be the first burden on the contractor – to establish an implied term that the engineer would carry out his duties under the contract. In the Sixth edition that burden is removed and the contractor has a much clearer entitlement to claim – breach of clause 2(1) (a).

Moreover, clause 2(1) (a) does not stop at requiring the engineer to carry out his duties specified in the contract: he must also carry out those necessarily to be implied from the contract. For example, clause 32 on fossils does not actually say that the engineer must give orders on their disposal but it certainly implies that he will do so if they are discovered.

The need for an express statement on the engineer's obligation to carry out his duties has long been apparent. Now that it is included in the Sixth edition engineers will have to be constantly vigilant that they, and anyone acting with delegated powers, do not, by any default of omission or delay, leave their employers liable for claims for breach.

Duties, authority and responsibility

Clause 2(1) (a) refers only to 'duties'. Other parts of clause 2 refer to the engineer's 'authority' and his 'responsibilities'. The word 'authority' does not appear anywhere in the Conditions outside clause 2 so it is necessary to consider how the engineer's authority arises and can be exercised.

A starting point is the proposition that the word 'shall' indicates a duty and the word 'may' indicates the power to exercise authority. Thus in clause 51, the engineer shall order any variation necessary to complete the works and may order any variation which is desirable for their completion.

This may be something of a simplification in that the exercise of a duty will frequently imply authority but it illustrates that there are two sources of authority. Firstly there is the contract; and if there are specified duties on a person, other than the parties, that person must be vested with the authority to perform those duties. In effect this source of authority comes jointly from the parties. Secondly there is the authority given to the engineer in his capacity as agent of the employer. This authority comes solely from the employer but it may be incorporated into the contract as a power. When the contract refers to authority it may encompass both the duty and the power or either.

The phrase 'responsibilities of the Engineer under the Contract' may be intended as no more than a collective phrase for duties and powers but there is the possibility it also takes in the matter of conduct in the way those duties and powers are to be exercised.

3.3 *Approval to exercise authority*

Clause 2(1) (b) of the Sixth edition is a new provision which states that the engineer may exercise the authority specified in or to be implied from the contract and then deals with the position where the engineer, under the terms of his engagement, needs to obtain the approval of the employer before exercising such authority.

This is recognition, long overdue, of the fact that an engineer rarely has unfettered power as agent of the employer. The most obvious matter is in the ordering of variations where the employer quite sensibly may impose financial limits on the engineer's spending power. There is nothing but trouble to be gained by the engineer ordering variations which the employer cannot afford. The variation clauses of the Fifth and Sixth editions, however, give no recognition to such reality and indeed, in stating that the engineer shall order any variation necessary to complete the works, seem to assume the employer has a bottomless purse.

The Sixth edition in clause 2(1) (b) has followed the route taken by the Fourth edition of the FIDIC Conditions published in 1987. It requires that particulars be set out in the appendix to the form of tender when the engineer, under the terms of his appointment, has to obtain the specific approval of the employer before exercising his authority under the contract.

Section 18 of the appendix reads:

> 'Requirement for prior approval by the Employer before the Engineer can act. DETAILS TO BE GIVEN AND CLAUSE NUMBER STATED
> ....
> ....'

Clearly this is intended to operate in a detailed way and a general statement of the engineer's duty of care to the employer will not suffice.

Matters to be included in the appendix

The question of what matters employers should include in the appendix has occupied much of the discussion time at seminars on the Sixth edition. Financial limitation on the power to order variations seems obvious enough as simple recognition of a practice already prevalent. But even for this there is some difficulty and for other matters there are serious doubts and complications.

Firstly there is a point of general principle on how the word 'authority' is to be interpreted in this context. Does the word relate to authority generally in the context of 'shall' and 'may' discussed above or is it confined only to the 'may' – that is to the exercise of discretionary power? If authority in this context does extend to 'shall' – that is, to the exercise of duties – to what extent is it proper or permissible for the employer alone to impose limitations on the engineer's authority?

Secondly, there is the complication introduced by the new sub-clause 2(8) on impartiality which states, in a curious piece of wording, that the engineer is to act impartially except in connection with matters requiring the specific approval of the employer under sub-clause 2(1)(b). This appears to be saying that the engineer need not act impartially on such matters. If this is the correct interpretation then such matters could not include, without making a nonsense of the whole balance of the contract, the authority of the engineer to exercise duties which demand impartiality, e.g.:

- granting extensions of time under clause 44
- valuing variations under clause 52
- making decisions under clause 66.

Authority in the context of clause 2(1)(b)

The simple answer appears to be that 'authority' in the context of clause 2(1)(b) should only relate to the exercise of discretionary power and not to duties. But consider the apparently obvious and seemingly proper limitation on the engineer's authority to order variations. Clause 51 contains both a duty, 51(1)(a) – (the engineer shall order any variation necessary for completion of the works) and a power, 51(2)(b) – (the engineer may order any variation desirable for the completion of the works). On the above analysis, only the latter should be included in the appendix but it is doubtful if this is the intended application of the clause.

Discretionary powers

What guidance then can be given on this provision? Certainly there should be little difficulty with exercise of discretionary powers such as:

(a) The site – clause 1(1)(v). Indeed the employer would be well advised to restrict the engineer's power to extend the site without approval.
(b) Engineer's representative and assistants – clauses 2(3), 2(4) and 2(5). The employer may want to have some say in appointments and delegation.
(c) Contractor's programme and methods – clause 14(2). The employer may not want the engineer to accept a programme showing completion in less than the contract time.
(d) Partial completion – clause 48(4). The employer may not want to take on prematurely the burden of taking completion of parts of the works he does not intend to use.
(e) Variations – clause 51(1)(b), discussed above.
(f) Daywork – clause 52(3). Some employers restrict, or try to limit, the ordering of dayworks.

But where authority does not relate to a discretionary power, its inclusion in the appendix will vary from being questionable to unsustainable depending on how far it curtails the independence of the engineer and impedes or prevents him fulfilling properly his duties under the contract.

Restrictions generally

Two further points should perhaps be added. One is that it would be unwise to restrict the freedom of the engineer to deal with safety matters – something which could arise unwittingly by requiring the engineer to seek approval before operating clause 40 (suspension of work). The other point is that clause 2(1)(b) deals only with the engineer's authority and it does not address the problem of audit interference. Such interference usually takes place outside the contract and not infrequently puts the employer in breach of it. Only if the engineer's authority to value and certify was curtailed could clause 2(1)(a) come into play as a breach.

Authority deemed to be given

The final sentence of clause 2(1)(b) says in effect that the engineer is deemed to have been given approval to exercise authority to the extent specified in the appendix.

It would surely be odd if it were otherwise and the contractor was put to the burden in every instance of enquiring whether the engineer had obtained the requisite approval. But at least by stating the point it avoids the possibility of delay in carrying out instructions, the costs which might fall on the employer, while proof of approval of authority was being sought.

No authority to amend terms and conditions

Clause 2(1)(c) states that the engineer has no authority to amend the terms and conditions of the contract, nor to relieve the contractor of any of his obligations except as stated in the contract.

Strictly, such a clause is unnecessary since the engineer derives all his authority to administer the contract from the contract itself and if the engineer has authority from other sources, perhaps in some statutory capacity, he does not exercise that authority in his role as the engineer.

The Fifth edition had no identical clause although clause 2(1) contained a similar statement of the obvious to the effect that the engineer's representative had no authority to relieve the contractor of any of his duties or obligations. Perhaps some assumed from this that the engineer did have such authority. However, whatever the reason for the inclusion of clause 2(1)(c) in the Sixth edition, it probably does no harm and it does bring home to the engineer and the parties a point which is sometimes overlooked.

There could even be cases where the express provision might usefully apply. Thus, if the engineer were to tell the contractor he would not suffer deduction of liquidated damages for delay this might be regarded as waiver by the employer – taking the engineer as agent of the employer. At common law the contractor might have a case but with this contractual statement the contractor's case would be seriously weakened if not lost.

3.4 *Engineer to be a named individual*

The Fifth edition came under criticism from the outset in that a firm could be named as the engineer but no individual had to be identified as the person responsible for carrying out the contractual duties. There was also some debate at that time, and it has not lessened since, on whether the engineer should have a hands-on role in contract administration or a detached role. The two issues are often seen as related but resolution of the first, by requiring the engineer to be a named individual, does not necessarily resolve the second, by making him an involved individual.

The first issue can be dealt with in the contract, and has been in the Sixth edition by clause 2(2) (a). The second issue is to some extent covered by the restrictions on delegation of the engineer's duties and authorities by the provisions of the contract but beyond that it would be difficult to regulate how the engineer should manage his affairs.

Engineer to be chartered

The main intention of clause 2(2) (a) at first seems clear: where the defined engineer is not a single named chartered engineer the contractor must be notified within seven days of the award of the contract of the named chartered engineer who will assume full responsibilities of the engineer under the contract. That is to say, if the defined engineer is a firm, a chartered engineer from that firm must be named to act as engineer.

Few would quarrel with that but is there more to the clause?

Putting aside the point that the reference to 'single' is presumably not intended to be to marital status, what is the position if the defined engineer is named but is not a chartered engineer? It certainly appears from strict reading of the clause that the named and defined engineer must nevertheless within seven days of the award of the contract notify the contractor of the chartered engineer who will act on his behalf.

This may be welcomed by chartered engineers and even by some contractors but it is more than likely that some employers, particularly those in the public sector, will find it wholly unacceptable. If, for example, it is the policy of a local authority that its director of technical services acts as engineer and reports to the council on the performance of contracts it is unlikely that they will accept or understand why the full responsibilities of the engineer should be given to a less senior member of the department. And what of the position of a local authority or other employer who does not have the services of a chartered engineer? Are they required to use another form of contract?

Named individual

Employers who wish to use the Sixth edition but who do not wish to be bound by the requirement for the engineer to be a chartered engineer can amend the conditions by deleting in clause 2(2)(a) the phrase 'a single named Chartered Engineer' and substituting 'an individual' and by deleting the later phrase 'the name of the Chartered Engineer' and substituting 'the name of the individual'. Additionally in clause 2(2)(b) the phrase 'named Chartered Engineer' would need to be deleted and substituted by 'named individual'. This, at least, has the virtue of bringing clause 2 into line with its marginal note, named individual.

Replacement of named individual

Clause 2(2)(b) is valuable in giving the engineer power to replace a named chartered engineer. Such power would probably be implied if the named individual died or possibly left the employment of the firm named as engineer but the implied term might not extend to a general freedom for the named individual to be replaced at the engineer's will.

3.5 Engineer's representative

The engineer's representative in both the Fifth and Sixth editions is intended to be the representative of the engineer on site. This was specified by the definition of engineer's representative in the Fifth edition and although the definition is open in the Sixth edition, the role given to him in clause 2(3)(b) is to watch and supervise the construction of the works.

The title of engineer's representative is therefore best carried by the resident engineer or similar rather than by some other assistant of the engineer. It is not necessary that the engineer's representative should be given any delegated power, nor is it necessary that any power which is delegated should be given to the engineer's representative (see comment on clause 2.4 below). The practice adopted in some quarters of naming an office-based project engineer as engineer's representative is therefore both inappropriate and unnecessary.

It is necessary under clause 2(3)(a) for the name of the engineer's representative to be notified to the contractor in writing and the same applies to other assistants under clauses 2(4) and 2(5). However, whereas for assistants the notice must also state duties and scope of authority, there is no requirement for this for the engineer's representative – unless, of course, he is to be given delegated powers.

Duties and responsibilities

The starting position of the engineer's representative is set out in clause 2(3)(b). He shall watch and supervise but he has no authority to relieve the contractor of any of his obligations. The clause then goes on to say that, except as expressly provided, he shall not order any work involving delay or extra payment nor make any variation in the works.

It is probable that the phrase 'except as expressly provided' in clause 2(3)(b) relates to the use of delegated powers but it may have a wider application to other clauses of the Conditions. However it must be said that there is very little in other clauses which could be taken as authorising the engineer's representative to order work or make variations.

Lack of express powers

In fact most clauses of the Conditions avoid reference to the engineer's representative and refer only to the engineer, thereby confirming that without delegated powers the role of the engineer's representative is no more than to watch and supervise. Even clauses 36, 37, 38 and 39 which come under the heading workmanship and materials refer only to the engineer, although it might be expected that if the phrase 'watch and supervise' was to confer any authority in its own right, it would be reflected in these clauses. Seemingly not – if the engineer's representative is to have any power under these clauses he can only acquire it through delegation under clause 2(4).

There is admittedly a statement in clause 13(1) that 'The Contractor shall take instructions only from the Engineer or (subject to the limitations referred to in Clause 2) from the Engineer's Representative'. However, it is not thought that this is intended to give power to the engineer's representative; it seems to be more by way of advice to the contractor not to take instructions unless there has been delegation.

A similar statement appears in clause 15(2) saying who the contractor's agent shall take directions and instructions from.

Minor powers

In clause 17 there is apparently a small departure from the policy of containing the duty of the engineer's representative to watching and supervising except where extended by delegation. Under this clause the employer is liable for the cost of rectifying setting-out errors resulting from incorrect data supplied by either the engineer or the engineer's representative. This implies that the engineer's representative has power to supply data on setting-out in his own right. Perhaps it is regarded as part of supervision.

Clause 19 on safety and security also includes reference to the engineer's representative in requiring the contractor to comply with instructions for the safety and convenience of the public or others from any competent authority. This again might be regarded as part of the duty of supervision.

The other few references in the Conditions to the engineer's representative relate to the receipt of information rather than the giving of instructions.

Reference on dissatisfaction

Under clause 2(7) if the contractor is dissatisfied with any act or instruction of the engineer's representative he is entitled to refer the matter to the engineer for his decision. This appears to apply to both the express powers of the engineer's representative and any delegated powers.

This reference is not intended to be a decision of the engineer under clause 66 but it may be one of the preliminary steps referred to in that clause to be taken prior to the serving of a notice of dispute.

3.6 Delegation by the engineer

Clause 2(4) of the Sixth edition permits the engineer to delegate any of his duties or authority except in respect of decisions or certificates under the eight clauses listed below:

clause 12(6) – determination of delay and extra cost for unforeseen conditions
clause 44 – assessment of delay and granting extensions of time
clause 46(3) – requests for accelerated completion
clause 48 – certification of substantial completion
clause 60(4) – certification of the final account
clause 61 – issue of the defects correction certificate
clause 63 – determination of the contractor's employment
clause 66 – engineer's decision following notice of dispute.

This is much the same as the position under clause 2(3) of the Fifth edition, but with the new clause 46(3) added.

There is however a wording change which may be of some administrative importance. The Fifth edition permitted delegation to be either generally in respect of the contract or specifically in respect of particular clauses. The Sixth edition is not clear on this point and it is possible that a notice of delegation has to be specific to be effective.

As mentioned earlier in this chapter, delegation can be either to the engineer's representative or to 'any other person responsible to the Engineer'. The phrase appears to give wide scope but it is also potentially restrictive since it would seem to exclude delegation to someone of equal status, for example, delegation between partners in a firm.

Under clause 2(4) (a) delegation must be in writing and it is not effective until notice is delivered to the contractor. The delegation remains effective, unless and until revoked in writing under clause 2(4) (b).

Nature of delegation

The wording of clause 2(4) suggests that the engineer can only delegate a duty or authority to one person. However the possibility that the engineer could, without breaching the provision, divide his duties and authorities and delegate amongst a number of persons cannot be totally ruled out.

There is in any event some difference of opinion on whether the engineer, having delegated a power, still has that power himself, or whether, at least until revoked, the power resides solely in the person to whom it has been delegated. The general rule that an authority which delegates its powers does not divest itself of them may be modified by particular provisions for delegation and one factor which may be relevant is the power to revoke. Other factors may be the circumstances of operation and the extent to which confusion would arise from the same power being exercised simultaneously by different persons.

If under the ICE Conditions the engineer does retain the powers he has delegated, and this is doubted, it does not diminish the effectiveness of actions taken by another person under delegated powers. These stand to be assessed and valued under the contract as if given by the engineer.

Restrictions on delegation

The restrictions on delegation imposed by clause 2(4)(c) cover some but not all of the crucial matters of the contract.

The engineer is still free to delegate on such important matters as:

(a) resolving ambiguities or discrepancies in documents under clause 5
(b) issuing instructions under clause 13
(c) suspending the progress of the works under clause 40
(d) valuing variations under clause 52.

Disputes on these can of course be referred back to the engineer under clause 2(7) and eventually clause 66, but in terms of the employer's liability to pay for action taken, the damage may already have been done.

Clearly engineers need to be mindful of their own professional and contractual obligations when considering on any contract which, if any, powers they should delegate.

One small point which disappointingly remains the same in the Sixth edition as in the Fifth is that although the engineer is unable to delegate his power to accept claims for unforeseen conditions under clause 12(6), he remains free to delegate his power to reject such claims under clause 12(5). Many would argue that this should be reversed.

3.7 *Assistants*

Assistant resident engineers and clerks of works play a vital part in the supervision of contracts. Clause 2(5) of the Sixth edition gives the engineer or the engineer's representative power to appoint any number of persons to assist the engineer's representative to carry out his duties of watching and supervising construction of the works or any other duties delegated to him. The Fifth edition allowed for the appointment of assistants only in respect of watching and supervising and, as will be seen below, the change may be of significance.

Under clause 2(5)(a) the contractor is to be notified of the names, duties and scope of authority of such persons. Presumably this should be in writing but it is not an express requirement. Perhaps the omission reflects the uncertainty as to the true contractual position of such persons – the Conditions recognise that they exist but do they really have a contractual, as opposed to practical, role?

Powers of assistants

The question is to what extent can assistants acquire duties and authority under the contract? Is it permissible for assistants to issue variations and instructions and to condemn materials and workmanship? Is the contractor obliged to comply with any instructions so given?

Under the Fifth edition the contractual role of assistants was confined to watching and supervising. It was generally held that they could not issue variations, nor could they give instructions on any matter outside the bounds of watching and supervising. Moreover it was strongly argued that, since their contractual power derived only from that specifically given to the engineer's representative in the Conditions and not to that be acquired by delegation, then they had no power in respect of clauses 36 to 39 on workmanship and materials. On this view, which was probably correct, an assistant could not order the contractor to send away materials thought to be outside specification; he could only advise the contractor and the engineer's representative of his findings.

Note, however, that under clause 2(5) (a) of the Sixth edition, assistants are specifically allowed to assist the engineer's representative in his delegated duties. Although clause 2(5) (b) says assistants have no authority to issue instructions save as necessary to enable them to carry out their duties and to secure their acceptance of materials and workmanship, they do have authority to issue instructions which are so necessary. Such instructions are to be in writing where appropriate and deemed to have been given by the engineer's representative.

It may follow from this that assistants can have the same duties and authorities as the engineer's representative has acquired by delegation if the engineer's representative notifies the contractor accordingly. There must be some doubt as to whether a change of this magnitude was intended. What is clear, having regard to the provision that any instructions given by an assistant in writing are deemed to have been given by the engineer's representative, is that the appointment of assistants under the contract needs to be handled with care.

Dissatisfaction with assistant's instructions

Under clause 2(5) (c) if the contractor is dissatisfied with any instruction of an assistant he is entitled to refer the matter to the engineer's representative who may confirm, reverse or vary such instruction. A similar provision existed in the Fifth edition.

3.8 Instructions to be in writing

The Sixth edition generally requires instructions to be in writing. The Fifth edition required only variations to be in writing.

Clause 2(5) (b) deals with instructions of assistants and allows some relaxation of the written requirement by the phrase 'where appropriate shall be in writing'. What is 'appropriate' will no doubt be a matter of debate.

Clause 2(6) (a) requires instructions given by the engineer or the engineer's representative exercising delegated powers to be in writing. Curiously the case of the engineer's representative acting under his own power is omitted. This has the odd result that an instruction of an assistant given in writing can be referred to the engineer's representative who can issue an oral instruction in reply. It was probably intended that all instructions by the engineer's representative should be in writing.

Clause 2(6) (a) further provides that the contractor shall comply with any 'such instruction' which it is necessary to give orally. The reference to 'such instruction' is unfortunate since the contractor apparently need only comply with oral instructions of the engineer's representative if given under delegated powers. The phrase 'any instruction' would have avoided the problem.

Confirmation of oral instructions

Clause 2(6) (b) extends to instructions a rule which previously only applied to variations – that if they are confirmed in writing by the contractor after being given orally they stand as instructions unless contradicted forthwith.

This should have a powerful effect in inducing the engineer, his representative and his assistants to put all instructions in writing. If they do not they must expect a flow of paperwork from the contractor which will need careful checking to avoid unintended financial liabilities falling on the employer.

As to the meaning of 'forthwith' the *Shorter Oxford Dictionary* says 'immediately, at once, without delay or interval'.

3.9 Specification of authority

Clause 2(6) (c) brings into the Sixth edition a provision which has been long awaited by contractors. Namely that on written request the engineer or engineer's representative shall specify in writing under which of his duties or authorities any instruction is given. This will not only impose discipline, it will also develop contractual awareness.

Engineers will have to be careful in the precision of their reply not to open the door inadvertently to claims – for example, by rejecting unsatisfactory work under clause 13 instead of under clause 39. The contractor has no right of payment under clause 39 but by imaginative argument he can develop a right under clause 13.

3.10 Impartiality

Clause 2(8) is a new provision requiring the engineer to act impartially within the terms of the contract, having regard to all the circumstances, except for matters requiring the specific approval of the employer. This latter point has been commented on earlier.

Many will question whether it is necessary or desirable to have a clause requiring the engineer to act impartially. He is under an implied duty to do so when adjudicating between the parties or issuing certificates. He is under no such duty when acting as agent of the employer. This clause seems to place a duty of impartiality on the engineer in all circumstances other than for matters specified in the appendix as requiring the employer's approval.

'Having regard to all the circumstances'

The phrase 'having regard to all the circumstances', which ends clause 2(8), is ambiguous. It may mean that the engineer must consider everything in his impartial approach or it may mean that, where the engineer is acting as agent of the employer, he need only consider the employer's interests.

Chapter 4

Assignment and sub-contracting

4.1 Introduction

This chapter deals with clauses 3 and 4 of the Sixth edition – assignment and sub-contracting. It does not cover the detailed provisions of clause 59 for nominated sub-contracting – those are dealt with in chapter 17. However, on the application of general principles it is worth noting here the wording of clause 59(3):

> 'Except as otherwise provided in Clause 58(3) [relating to design] the Contractor shall be responsible for the work executed or goods materials or services supplied by a Nominated Sub-Contractor employed by him as if he had himself executed such work or supplied such goods materials or services.'

The substance of clause 3 remains unchanged except that the employer and contractor are now both subject to the clause. In the Fifth edition it applied only to the contractor.

The changes to clause 4, giving the contractor the right to sub-contract without any need to obtain the approval of the engineer, represent a major policy shift. Some engineers and employers have indicated they find the changes an unacceptable erosion of their control over the construction of the works.

4.2 Assignment and sub-contracting distinguished

Definitions

Assignment can be simply defined as a transfer of title or interest in land, property or contractual rights.

Sub-contracting can be defined as an arrangement to secure vicarious performance of contractual obligations.

To the practical eye the end result may often seem to be the same but in law the treatment of rights and obligations is different. And assignment and sub-contracting are quite different in their approach to the basic doctrine of privity of contract. Assignment is an exception to the doctrine of privity; sub-contracting gives no relief from the doctrine.

Assignment then is a process which confers legal rights and can include the transfer but not extinguishment of contractual obligations. Sub-contracting is no more than delegation of performance. It does not relieve the contractor of his obligations to the employer and it does not create any contractual rights or obligations between the employer and sub-contractor.

4.3 *Doctrine of privity of contract*

The essence of the doctrine of privity of contract is that only the parties to the contract acquire rights and obligations under the contract. This applies even to contracts made with the intention of conferring rights on a third party as illustrated by the case of *Beswick* v. *Beswick* (1968) where a widow was unable to sue her nephew for payments he had promised her husband he would make to her as part of the price of acquiring his business.

The Lord Chancellor, Viscount Haldane expressed the doctrine of privity in this way in *Dunlop* v. *Selfridge* (1915):

> 'My Lords, in the law of England certain principles are fundamental. One is that only a person who is a party to a contract can sue on it. Our law knows nothing of a *jus quaesitum tertio* arising by way of contract. Such a right may be conferred by way of property, as, for example, under a trust, but it cannot be conferred on a stranger to a contract as a right to enforce the contract *in personam*.'

In that case Dunlop were attempting to sue a retailer for breach of a retail price agreement but Dunlop only had a contract with the wholesaler who was not himself in breach.

Chain of responsibility

One effect of the doctrine of privity of contract is to establish what is known as a contractual chain of responsibility. If a party at the bottom of the chain in breach of contract causes the party above him to suffer damage and this is transmitted upwards through the chain, recovery of damages at the higher levels can only be achieved by a series of actions. Leapfrogging is not permitted even when a party has gone out of business and there must be continuity of breach and damage at each link. In the *Dunlop* case above there was no continuity of breach and probably no continuity of loss since the wholesaler may have gained rather than lost by the retailer's breach.

Collateral contracts

There is a device, applicable only in certain circumstances, by which a party can seek to bridge a gap in a contractual chain by establishing what is termed a collateral contract. Such a contract must contain the essentials of offer, acceptance and consideration and it would usually be not far removed from a guarantee.

In the well known case of *Shanklin Pier Ltd* v. *Detel Products Ltd* (1951) the contractor was instructed to use paint supplied by Detel on the basis of a representation made by Detel that the paint would last for seven years. When the paint failed the employer, Shanklin Pier, successfully sued the supplier, Detel, on the basis that consideration lay in specifying the paint in return for the promise of its durability.

Collateral warranties

These operate on a different basis than collateral contracts. Collateral warranties are undertakings, formally executed, to create direct contractual relationships where none would otherwise exist. Their purpose is to create third party rights over and above those available in tort where, with limited exception, the duty of care no longer extends to the prevention of economic loss.

4.4 Assignment and novation

Benefits and burdens

As a general rule, English law permits the assignment of rights but not of obligations. The position is usually explained in terms of benefits and burdens as here by Lord Justice Bingham in the case of *Southway Group Ltd* v. *Wolff and Wolff* (1991). He said:

> 'It is in general permissible for A, who has entered into a contract with B, to assign the benefit of that contract to C. This does not require the consent of B, since in the ordinary way it does not matter to B whether the benefit of the contract is enjoyed by A or by a third party of A's choice such as C. But it is elementary law that A cannot without the consent of B assign the burden of the contract to C, because B has contracted for performance by A and he cannot be required against his will to accept performance by C or anyone other than A. If A wishes to assign the burden of the contract to C he must obtain the consent of B, upon which the contract is novated by the substitution of C for A as a contracting party.'

The position in a construction contract is that the employer has the benefit of the finished works and the burden of paying for them. The contractor has the burden of constructing the works and the benefit of being paid.

Novation

A novation is a tripartite agreement whereby a contract is rescinded in consideration of a new contract being entered into, on the same or similar terms as the old contract, by one of the original parties and a third party. This frequently occurs when one of the original parties changes its legal status or goes into receivership. Novation overcomes the doctrine of privity of contract and the restrictions the law places on assignment.

Novation has the effect of discharging the parties from the obligations of the original contract. It is not always appreciated that assignment does not achieve this. Thus where a lease to a property is assigned the original lease holder remains liable to the landlord for payments due until the lease expires.

Restrictions on assignments

The general rule that the benefits of a contract can be assigned simply by notice but the burdens can only be assigned with the consent of the other party is subject to some important exceptions.

Two of these are relevant to the construction industry.

The first is that the contract itself may expressly provide that neither rights nor benefits shall be assigned. Since this is the case in the ICE Conditions, the background and significance of the point is discussed in more detail later in this chapter.

The second exception, which has its origins in the development of common law and equity, is that the law does not recognise assignment of the mere right to sue for damages – in legal terms, a bare right of action – unless the assignee has a genuine commercial or financial interest in the assignment.

This does not invalidate the practice of factoring debts – which can be assigned – but it prevents the passing on for monetary gain of the right to sue on a breach. Thus a contractor could not, after completing the works, sell on his rights for unsettled claims. However, the courts recognise that in commercial transactions there may be genuine reasons for transferring rights of action.

In the case of *South East Thames Regional Health Authority* v. *Lovell and Others* (1985) one of the defendants, a firm of structural engineers, acquired under terms of a settlement the right to continue the action in the names of other defendants. It was held that the settlement agreement included an assignment of rights of action but this was not void for champerty or maintenance since the structural engineers had a genuine commercial interest in the transaction notwithstanding the fact that they might make a profit out of the litigation. The common law misdemeanour of champerty was abolished by the Criminal Law Act 1967, but it is still against public policy to support an action in return for a proportion of the amount awarded in proceedings.

4.5 *Nature of sub-contracting*

The courts have frequently had to consider whether the sub-contracting of obligations amounts to assignment. This is how the Master of the Rolls, Lord Greene, considered the matter in *Davies* v. *Collins* (1945). He said:

> 'In many contracts all that is stipulated for is that the work shall be done and the actual hand to do it need not be that of the contracting party himself; the other party will be bound to accept performance carried out by somebody else. The contracting party, of course, is the only party who remains liable. He cannot assign his liability to a sub-contractor, but his

liability in those cases is to see that the work is done, and if it is not properly done he is liable. It is quite a mistake to regard that as an assignment of the contract; it is not.'

Wrongful sub-contracting

The implications of wrongful sub-contracting, therefore, are not drawn from the rules on assignment but are a separate issue. They are concerned not with whether the sub-contractor has a right to sue the employer for payment (clearly he has not) but whether the employer can refuse to pay the contractor for work he has sub-let.

Leaving aside express provisions in contracts which may or may not permit sub-contracting and may or may not state the remedies for breach, what is the general position on this?

The courts apply the test that where a person is selected with reference to his individual skill, competency or qualifications, he cannot sub-contract performance. To do so would be a breach of contract which could jeopardise his right to payment.

Thus in the *Southway Group* case mentioned above in 4.4 the Court of Appeal held that personal performance by the contractor was essential since the employer had selected the contractor for his particular skills. The employer was not obliged to pay for substituted performance.

The test for permissible sub-contracting is whether, from the terms of the contract and its circumstances, it can be inferred that it is a matter of indifference whether performance is sub-contracted or not.

4.6 *Contractor's responsibility for sub-contractor's performance*

The general principle is that the contractor is not relieved of any of his obligations under the contract by sub-contracting. But this is not to say that the contractor is fully responsible for the performance of his sub-contractor in the wider legal context of statute and tort. And even in contract, relief may sometimes be given to the contractor from his strict obligations by express provisions of the contract – not all of which, it must be said, are necessarily drafted with that end in mind.

Responsibility in tort/duty of care

The contractor's responsibility for defective work undertaken by a sub-contractor is different in tort than in contract. In *D.F. Estates Ltd* v. *Church Commissioners for England* (1988) it was held that the contractor discharged his duty of care by appointing a sub-contractor he believed to be reasonably competent and any duty to supervise the sub-contractor was solely in contract.

'Beyond the contractor's control'

This phrase was given unexpectedly wide meaning by the House of Lords in the case of *Scott Lithgow Ltd* v. *Secretary of State for Defence* (1989). It was ruled that the contractor was entitled to payment under a provision which included, amongst other things, as grounds for extra payment the phrase 'or any other cause beyond the contractor's control'. Lord Keith said:

> 'Failures by such suppliers or sub-contractors, in breach of their contractual obligations to Scott Lithgow, are not matters which, according to the ordinary use of language, can be regarded as within Scott Lithgow's control.'

The case concerned nominated specialist suppliers and sub-contractors but it may nevertheless have wider application. Accordingly many standard forms of contract which include the phrase 'beyond the Contractor's control' in relation to payment, extension of time or other matters have been amended.

The JCT Minor Works Agreement, for example, which used the phrase in its extension of time clause, now has an extra sentence:

> 'Reasons within the control of the Contractor include any default of the Contractor or of others employed or engaged by or under him or in connection with the Works and of any supplier of goods or materials for the Works.'

Restricted interpretation of 'contractor'

On general principles one would not expect the meaning of 'contractor', when used in the provisions of a standard form, to exclude 'sub-contractors'. However in *Jarvis Ltd* v. *Rockdale Housing*

Association Ltd (1986) the Court of Appeal held that 'contractor' in the phrase 'unless caused by some negligence or default of the contractor' in the determination provisions of JCT 80 applied only to the contractor as main contractor and did not include nominated sub-contractors.

These two extracts from the judgment of Lord Justice Bingham explain the logic of the decision:

> 'I cannot accept the employer's submission that "the Contractor" in 28.1.3.4 is to be understood as including sub-contractors and their servants and agents. Both in this clause and elsewhere in the contract the draftsman has made express reference to sub-contractors, their servants and agents. The absence of any such reference here is not explained by the context, nor can it be dismissed as of no significance. Even an inexperienced draftsman could be relied upon to avoid such an error.'
>
> "The Contractor" can in my judgment be naturally and sensibly understood as referring to, in this case, John Jarvis Ltd, its servants and agents, through whom alone it can, as a corporation, act.'

It is clearly most important that contract draftsmen express their intentions with both clarity and consistency. The old maxim 'never change your wording unless you intend to change your meaning' is evidently as good as ever.

4.7 Assignment in the Sixth edition

Clause 3 of the Sixth edition is broadly the same as in the Fifth edition except that it now applies with equal force to both the employer and contractor. It additionally says that the consent which is required of the other party to assignment 'shall not unreasonably be withheld'.

Assignment of the contract

Neither party has a right to assign 'the Contract' at common law so the statement in clause 3 does not add anything by way of prohibition.

However it could be argued that the concluding words in the clause on consent not being unreasonably withheld imply some entitlement to assign the contract. If such an argument were to succeed the contract would be gravely flawed.

It is probable that the words are intended to apply only to the assignment of benefits or interests because in the ordinary course of things their assignment does not require consent.

Assignment of benefit or interest

Clause 3 prohibits the assignment of any benefit or interest without prior consent.

Many standard forms of contract are silent on this, the parties being satisfied, one presumes, with the assignment of benefits and interests in accordance with common law and equity. Why then should the parties wish to prohibit the assignment of benefits and interests?

Firstly, there is the position of the employer in restraining the contractor from assigning benefits. On this there is some judicial comment, albeit under the ICE Fourth edition. This is how Mr Justice Croom Johnson put the matter in *Helstan Securities Ltd* v. *Hertfordshire County Council* (1978):

> 'The clause is obviously there to let the employer retain control of who does the work. Condition 4, which deals with sub-letting, has the same object. But closely associated with the right to control who does the work is the right at the end of the day to balance claims for money due on the one hand against counterclaims, for example, for bad workmanship on the other. The plaintiffs say that such a counterclaim may be made against the assignees instead of against the assignors. But the debtors may only use it as a shield by way of set-off and cannot enforce it against the assignees if it is greater than the amount of the debt: *Young* v. *Kitchin* (1878). And why should they have to make it against people whom they may not want to make it against, in circumstances not of their choosing, when they have contracted that they shall not?'

Then there is the position of the contractor restraining the employer. In this the issue will usually be the employer's rights in connection with remedial works and latent damage. Is it appropriate that these should be assigned, perhaps in association with the transfer of property, with the result that the contractor might be sued by an occupier or user of the works he had not contracted with and never contemplated doing business with?

The legal issues here are complex, as seems inevitable with the transfer of property, but the following points can be drawn from decisions of the Court of Appeal in two building cases which prohibited assignment of 'the Contract' – *Linden Gardens Trust Ltd* v. *Lenesta Sludge Disposals Ltd* (1992) and *St Martin's Property Corporation Ltd* v. *Sir Robert McAlpine & Sons Ltd* (1992):

(a) Prohibition of assignment of 'the Contract' does not, itself, preclude assignment of benefits arising 'under' the contract.
(b) Express words are needed to prohibit the assignment of benefits.
(c) If a cause of action arises before assignment this can be a benefit 'under' the contract.
(d) If there is no cause of action before assignment, then there is only 'the Contract' to assign.

The Sixth edition, by its comprehensive wording, at least avoids dispute on how 'the Contract' and 'benefits under the Contract' differ since assignment of both is expressly prohibited.

Is a debt a benefit?

Prior to the *Helstan* case mentioned above many legal commentators believed that assignment of a debt due was not caught by contractual restrictions on the assignment of benefits. There is still some uncertainty on the point and more so on whether it is actually intended that the restrictions on assignment of benefits should include charges on money due. Some standard forms avoid the uncertainty by adding that such charges shall not be considered assignments.

Assignment without consent

The sanction for the employer in the Sixth edition if the contractor assigns 'the Contract' without consent is determination of the contractor's employment under clause 63(1) (a) (ii).

It is doubtful if this covers the mere assignment of benefits since the clause 63 wording does not follow exactly that in clause 3.

The contractor has no express sanctions for the employer's

unauthorised assignment but he may have a stronger common law case to determine for breach because of this.

4.8 Sub-contracting under the Sixth edition

Clause 4 of the Sixth edition represents a major change from the Fifth edition in generally freeing the contractor from restrictions on sub-contracting. This is acceptance of the fact that the majority of work in civil engineering is undertaken by sub-contractors.

Not everyone has accepted that the change was necessary or desirable and some engineers and employers have indicated that they will amend the Conditions to retain a degree of control over the employment of sub-contractors by re-introducing consent measures. There is certainly some force in the argument that, where an employer operates a select list of contractors based, amongst other things, on past satisfactory performance, he would not wish to see a struck-off contractor re-emerge as the major sub-contractor for one of his contracts, undertaking perhaps 90% or more of the works.

But against this it can be said that the employer and engineer can, and should, deal only with the contractor; that in strict contractual terms the contractor should be free to undertake the performance of his obligations as he sees fit; and that the involvement of the employer or the engineer in the contractor's arrangements gives unnecessary potential for dispute.

It is certainly true that clause 4 of the Fifth edition generated a lot of friction and was open to various interpretations on how the consent provisions applied to individual sub-contractors, as opposed to the work they were to undertake. Contractors were offended to be told they could not sub-let to the firm which they preferred, and indeed might have relied on in their tender; and engineers were never certain how to cope with the unauthorised sub-contracting which inevitably took place. There was a draconian power of forfeiture under clause 63 for sub-contracting in defiance of the engineer but it was rarely used.

On balance it is probably best that the contractor has freedom of action.

The whole of the works

Clause 4(1) states that the contractor shall not sub-let the whole of the works without the prior written consent of the employer.

There is no reference, as in clause 3, to this consent not being unreasonably withheld so it is absolutely at the discretion of the employer how he deals with any request. Note that it is the consent of the employer which must be given – not that of the engineer.

There is some question as to what is meant by 'the whole of the Works'. Does it apply only to the whole of the works as a single sub-contract package or does it apply where the total of sub-contracted work amounts to the whole?

The answer is probably both but it would seem there is nothing to stop the contractor sub-contracting just short of the whole. Whether the retention of only a supervisory role, while sub-letting all the physical work, would amount to breach is a matter of debate.

Any part of the works

Clause 4(2) expressly permits the contractor to sub-contract 'any part of the Works' or their design. The only stipulation is that the extent of the work sub-contracted and the name and address of the sub-contractor must be notified in writing to the engineer prior to entry on site or appointment in the case of design.

It has been suggested by some commentators that the opening words of clause 4(2) 'Except where otherwise provided' are a device whereby the contractor's freedom to sub-contract without approval can be overridden in some other contract document. Perhaps this is correct but it is difficult to see how any amendment would work without deletion of the clause. It is more logical to consider the words to be a reference to the provisions of clause 59 on nominated sub-contracting.

Failure to give notice

There is no express sanction in the Conditions for failing to give notice of sub-contracting and the Sixth edition does not contain any specific reference in clause 63 (determination) comparable to that in the Fifth edition. It is most unlikely that the employer could refuse to pay for any work executed to the required standard or that the employer could prove any damage from the breach of failing to give notice.

The general provision in clause 63 – 'persistently or fundamentally in breach of his obligations' – might apply in extreme cases of wilful refusal to give notice of, or to name, sub-contractors.

Labour-only sub-contractors

Clause 4(3) confirms, as did the Fifth edition, that labour-only sub-contractors need not be notified.

Under the Fifth edition this was described as labour on a piece-work basis but, of course, in practice many labour-only sub-contractors work on a daywork basis.

Contractor liable for sub-contractor's work performance

Clause 4(4) states that the contractor is liable under the contract for the work, acts, defaults and neglects of his sub-contractors, agents, servants or workpeople.

This may not be intended as anything more than a reminder of the ordinary legal position but by a slight change of wording from the Fifth edition it may have greater significance. The question is, do the words 'liable under the Contract' cover acts of tort by a sub-contractor as well as contractual defaults?

The answer is probably, yes – but only in relation to dealings between the employer and the contractor. That is to say, the clause would not make the contractor liable to third parties for sub-contractor's torts.

Removal of sub-contractors

Clause 4(5) empowers the engineer to order the removal of any sub-contractor for misconduct, incompetence, negligence or breaches of health and safety rules.

This is an additional power to that given in clause 16 for the removal of the contractor's employees for similar conduct. Clause 16 clearly relates to individuals whereas clause 4(5) could cover firms.

If an engineer is in doubt as to whether clause 16 covers the individual employees of a sub-contract firm he would do well to give notice under both clauses although the procedure under clause 4(5) is more precise.

Under clause 4(5) the engineer must first give a warning in writing. He can then serve notice of removal – presumably again in writing but not stated. No time limits are set, as in clause 63 on the determination of the contractor's employment, but this is sensible in the light of safety considerations.

It is suggested that failure by the contractor to comply with an order for removal would be a serious breach entitling the engineer to consider action under either clause 40 (suspension) or clause 63 (determination).

Chapter 5

Contract documents

5.1 *Introduction*

This chapter covers:

clause 5 – documents mutually explanatory
clause 6 – supply of documents
clause 7 – further drawings and instructions.

The changes from the Fifth edition are minor although the procedures in clause 7 for parts of the permanent works designed by the contractor are new.

5.2 *Clause 5 – documents mutually explanatory*

Clause 5 in the Sixth edition is identical to that in the Fifth. This is one of the few clauses with no change whatsoever in wording or arrangement.

The clause states that the documents forming the contract are to be taken as mutually explanatory and any ambiguities or discrepancies shall be explained and adjusted by the engineer who shall thereupon issue instructions in writing to the contractor. Such instructions are to be regarded as instructions issued under clause 13.

Scope and purpose of clause 5

Since its introduction in 1973, clause 5 has troubled lawyers on its scope and purpose. It is argued that the engineer can only issue instructions relating to the construction of the physical works and he is therefore restricted to dealing with ambiguities or discrepancies of a technical nature. It is further argued that there is nothing in clause 5 which entitles the contractor to payment for errors in documentation which may have misled him in the compilation of his tender but are themselves errors which can be resolved without instructions.

And it is then said with some force that clause 5 is not a vehicle for the engineer to modify or alter the terms and conditions of the contract which the parties have themselves agreed. Nor, it is said, does clause 5 give the engineer any power to interpret those terms and conditions.

Engineers who have lived happily with clause 5 for 20 years may find all this surprising but clearly it is worth taking a look at the wording to see what difficulties can arise.

Documents forming the contract

These are defined in clause 1(e) as:

- Conditions of contract
- Specification
- Drawings
- Bill of quantities – priced and completed
- Tender
- Written acceptance of tender
- Contract agreement (if completed).

Clauses 1(f) and 1(g) make it clear that specifications and drawings can be modified and added to from time to time by the engineer so clause 5 is not restricted, for those classes of documents, to documents in issue at the time the contract was made. There is no general provision for change to other classes of documents and any change is a matter of agreement between the parties.

Documents mutually explanatory

There is little doubt that one purpose of the statement in clause 5 that the documents forming the contract are to be taken as mutually explanatory of one another is to avoid the difficulties which arise when there is an order of precedence. This has the effect, at least for the purposes of this clause, that the usual rule that special provisions prevail over standard provisions does not apply. The engineer can then explain and adjust ambiguities or discrepancies to the best practical solution unhampered by legal niceties.

However, it does not follow that clause 5 has the effect of reducing all documents to equal status for the purpose of interpretation of legal as opposed to practical issues. For example, if the appendix to the form of tender stated liquidated damages at £10,000 per week but the contractor in his tender said he would pay only £1000 per week and the tender was accepted, it is unlikely that clause 5 would be of much assistance in reaching a decision on which rate should apply.

This example also illustrates the restricted scope of clause 5. Surely it should not fall to the engineer to deal with such a matter under this clause. The better solution would be a reference to the engineer under clause 66 followed by conciliation or arbitration.

The words 'mutually explanatory' may have effect in cases on whether or not payment is allowable or a variation order is necessary, or can be deemed to have been given, following an instruction under clause 5. If work is shown on one drawing but not on another, the engineer is obliged to decide what is to be done. But work shown on a drawing will not normally be a variation; and in considering whether or not the contractor is entitled to extra payment under clause 13(3) the question of the contractor's obligation to take the documents as 'mutually explanatory' will be relevant.

Ambiguities or discrepancies

It is only ambiguities or discrepancies which are to be explained and adjusted by the engineer under clause 5. The engineer does not have any power under this clause to rectify errors in documents which are clearly expressed but which may nevertheless be the subject of dispute between the parties.

The words 'ambiguities' and 'discrepancies' have no precise legal meaning and their ordinary meanings – 'open to more than one interpretation' and 'difference' – do not include straightforward omissions. The engineer would have to look to his powers under clauses 13 or 51 to deal with these.

Engineer to explain and adjust

The requirement is that the engineer 'shall' explain and adjust ambiguities or discrepancies. This is a duty and not the exercise of discretionary power.

The clause does not demand an application by the contractor prior to the engineer's action and although in practice it will normally be the contractor who raises an issue, it would seem to be the duty of the engineer to act on his own initiative as soon as he becomes aware of any relevant issue.

In fulfilling his duty of explaining and adjusting, the engineer must principally have regard to his obligations under clauses 13 and 51 to secure the construction and completion of the works. Where he is doing this, considerations of fairness and impartiality will not apply. The engineer must simply decide what is required as a practical measure.

If, as some argue, this practical approach is the full extent of the application of clause 5 the engineer need not concern himself with the rules which the courts would apply in reaching their decisions – the rules of construction as they are called. However, if the engineer is confronted with a problem under clause 5 which, instead of having a practical solution requires him to be fair and impartial, he would be wise to adopt the same rules.

Rules of construction

These are briefly:

(a) Intention to be found from the contract itself. The courts will not go outside the written documents and substitute the presumed intention of the parties.
(b) Words are to be given their ordinary or plain meaning. In trade or technical contracts the customary meaning applies.
(c) Words are construed to make a contract valid rather than invalid.
(d) The intention of the parties is to be derived by construing a contract as a whole.
(e) Unless the documents expressly provide otherwise, particular conditions prevail over standard conditions.
(f) *Expressio unius* – express inclusion of a certain thing excludes others of a similar nature.
(g) *Eiusdem generis* – when words of a particular class are followed by general words, the general words are taken to apply to things of the same class.
(h) *Contra proferentem* – where there is ambiguity in a document

the words are to be construed against the party who put forward the document.

Instructions to be regarded as 'instructions in accordance with clause 13'

The provision that instructions given under clause 5 are to be regarded as instructions under clause 13 gives strong support to the argument that clause 5 is of practical application only.

Clause 13 requires the contractor to construct and complete the works strictly to the engineer's instructions. Although the wording is not itself restrictive on the range of instructions that the engineer can issue, the intention is clear enough. Moreover it is significant that reference in clause 13(3) back to clause 5 provides the only express entitlement to payment for a clause 5 instruction.

Instructions to the contractor

Finally there is the point that instructions are to be given 'to the contractor'.

If it had been intended that clause 5 should provide a wider power than suggested for the engineer to resolve ambiguities and discrepancies it is unlikely that it would have excluded the employer from its wording.

Use of clause 5

On balance the arguments come down in favour of clause 5 being restricted to practical measures and against its application to wider matters of legal interpretation.

5.3 *Clause 6 – supply of documents*

A few changes in detail have been made between the Fifth and Sixth editions on the supply of documents.

Number of copies

Clause 6 of the Sixth edition entitles the contractor to four free copies of the conditions of contract, specification and unpriced bill of quantities. The Fifth edition allowed only two copies of each.

For drawings, there is a new requirement in the Sixth edition that the number of copies to be provided free of charge shall be entered in section 4 of the appendix to the form of tender. In the Fifth edition, two free copies were provided.

Note that the contractor becomes entitled to his free copies of documents 'upon award of the contract' and these copies must be 'furnished'. This suggests that telling the contractor he must make do with any tender documents he was entitled to retain after submission of his tender is not intended and will not suffice.

Contractor's design drawings

Clause 6(2) requires the contractor to supply to the engineer four copies of all drawings, specifications and documents relating to any permanent works designed by the contractor and approved by the engineer under clause 7(6).

The contractor must supply further copies at the employer's expense if the engineer so requests in writing.

These are new provisions reflecting the incorporation in the Sixth edition of procedures for contractor's design of part of the permanent works.

Copyright and reproduction

Clause 6(3) confirms that copyright of drawings, specifications and bills of quantities does not pass to the contractor, nor does copyright of documents supplied by the contractor pass to the employer. However, the contractor is given freedom to reproduce for 'the purposes of the contract' and the employer and engineer can reproduce for 'the purpose of completing operating maintaining and adjusting the Works'.

The difference in phraseology is necessary. The contractor's role is to construct and complete the works. The employer and engineer are involved in long term operation and maintenance.

Return of documents

The provision in the Fifth edition that the contractor should return on completion of the contract all copies of drawings and specifications, whether supplied or reproduced, has been omitted from the Sixth.

This may be in recognition of the practical difficulties involved which led to widespread disregard. Alternatively the omission may just be to keep the positions of the contractor and employer in balance.

5.4 *Clause 7 – further drawings and specifications*

Clause 7 of the Sixth edition is expanded to cover drawings and specifications relating to contractor's design. The responsibility of the engineer to provide further drawings and specification and the contractor's rights to claim for delay in issue remain broadly as in the Fifth edition.

Engineer to supply further drawings etc.

Clause 7(1) places a duty on the engineer to supply such modified or further drawings, specification and instructions as are, in his opinion, necessary for the construction and completion of the works.

This is consistent with the engineer's duty to order under clause 51 any variation which is necessary for the completion of the works. Indeed clause 7(1) confirms that any drawings, specification or instructions which require a variation shall be deemed to have been issued pursuant to clause 51.

The express linkage between clause 7 and clause 51 suggests that work arising from modified or further drawings, specifications and instructions should be valued in accordance with the principles for the valuation of variations. However, note that clause 52(3) allows additional or substituted work to be executed on a daywork basis if it is necessary or desirable in the opinion of the engineer.

Extent of engineer's duty

When the Fifth edition came out there was some concern that the engineer was placed under a 'duty' to supply additional drawings etc. It was suggested that this weakened the contractor's basic obligation to construct and complete the works and placed a duty on the engineer to act when the contractor got into difficulties. Taken together with the obligations on the engineer under clauses 13 and 51, there is certainly some scope for dispute. But since such dispute is likely to be about money it is more likely to be argued out under either clause 13 or clause 51 than clause 7.

Moreover, in both Fifth and Sixth editions the test for action in clause 7 is 'as shall in the engineer's opinion be necessary' and except to the extent this may not allow much flexibility in the case of a variation, it is not as firm a duty as under clause 51.

Some recent concern has been expressed that clause 7(7) of the Sixth edition, which makes the engineer responsible for integration and co-ordination of the contractor's design, has widened the scope of the engineer's duty under clause 7(1) to supply further drawings etc. It may well be that there will be problems at the interface of the engineer's design and the contractor's design but the engineer is not necessarily burdened with the full responsibility for resolving this. See the comment on clause 7(2) below.

A commonplace practical problem under the Fifth edition, which remains to be resolved under the Sixth edition, is where the contractor presses the engineer for working drawings of construction details. Civil engineering contractors are generally supplied with more detail than building contractors but they cannot expect to have every nut and bolt and every level detailed however convenient that might be. It comes back to what in the engineer's opinion is necessary.

'From time to time'

The phrase in clause 7(1), requiring the engineer to supply further drawings etc. 'from time to time', indicates that it is permissible to supply information as the works proceed. It would not be possible in the light of this clause for the contractor to mount a claim alleging breach of contract for failure by the engineer to supply all information on commencement.

'During the progress of the works'

It was thought that under the Fifth edition the phrase 'during the progress of the Works' prevented the engineer from issuing further drawings etc. during what was then termed the maintenance period.

The phrase may not have the same effect in the Sixth edition since clause 51 has been amended to permit the issue of variations during the defects correction period and clearly any such variation might require the engineer to supply further drawings etc.

Contractor to provide further documents

Clause 7(2) entitles the engineer to request from the contractor further documents relating to parts of the works designed by the contractor.

The engineer would be well advised, having regard to his obligations under clause 7(7), to ensure that any parts of the works so designed can be integrated into the rest of the works and that he is provided with sufficient details to be certain of this.

Notice by contractor

Clause 7(3) requires the contractor to give adequate notice in writing to the engineer of any further drawings etc. which he requires for the construction and completion of the works.

This is not to say that the engineer's duty to supply further drawings etc. under clause 7(1) is only activated by the contractor's notice. It is clearly not since it is expressed as an independent duty. However, failure by the contractor would be relevant, but not decisive, to the matter of delay in issue discussed below.

As to what is adequate notice, the building case of *London Borough of Merton* v. *Leach* (1985), although not strictly comparable, is instructive. The contractor had set out in his programme the dates by which all information was required. It was held that an application might be made at commencement for all the instructions the contractor could foresee, provided the dates specified met the contractual requirements of being not unreasonably distant from nor unreasonably close to the relevant dates.

5.5 Delay in issue of further drawings etc.

By clause 7(4) the contractor is given an entitlement in certain circumstances to claim an extension of time and recovery of extra cost if delay is suffered as a result of the late issue of further drawings etc.

The Fifth edition contained a similar provision and the only change is that a new sub-clause 7(4)(b) restricts the contractor's right where he has contributed to the delay by himself not submitting in good time to the engineer details of contractor-designed parts of the works.

Application of clause 7(4)

Note particularly that the contractor's rights of claim under clause 7(4) relate only to drawings etc. 'requested by the contractor'.

The contractor cannot claim under this clause for delays caused by the issue or late issue of drawings etc. which the engineer issues on his own initiative under clause 7(1). Where such drawings etc. are issued in connection with a variation, all cost, including any delay which the contractor can substantiate, should be included in the valuation of the variations under clause 52. Where the drawings etc. are not issued in connection with a variation, but are perhaps no more than fuller working details, the best the contractor can do by way of a contractual claim if there is delay is to look to his rights under clause 13 – cost which could not have been foreseen at the time of tender.

'At a time reasonable in all the circumstances'

Clause 7(4)(a) limits the contractor's right of claim to delay caused by any failure or inability of the engineer to issue 'at a time reasonable in all the circumstances' the further drawings etc. requested by the contractor.

The question of what is reasonable can only be a matter of fact to be decided on the circumstances of each case. Amongst the circumstances to be considered are:

(a) provisions of the contract
(b) nature of the work
(c) custom in the industry
(d) resources of the parties and the engineer
(e) progress of the works.

The list is not conclusive and is merely an indication of matters put forward when disputes arise.

The limited judicial guidance which is available is no more than illustrative because each case turns on its particular merits but two civil engineering cases, both with provisions similar to those in the ICE Fifth and Sixth editions, are worth mentioning.

In *Neodox Ltd* v. *Swinton and Pendlebury Borough Council* (1958) the court was asked to decide whether there was an implied term in the contract that the engineer would give all details and instructions necessary for the execution of the works in sufficient time for the contractor to execute and complete the works in an economic and expeditious manner and/or in sufficient time to prevent the contractor from being delayed. In holding that it was impossible to imply such a term in the manner framed, Mr Justice Diplock said this:

> 'It is clear from these clauses which I have read that to give business efficacy to the contract, details and instructions necessary for the execution of the works must be given by the engineer from time to time in the course of the contract and must be given in reasonable time. In giving such instructions, the engineer is acting as agent for his principals, the Corporation, and if he fails to give such instructions within a reasonable time, the Corporation are liable in damages for breach of contract.
>
> What is a reasonable time does not depend solely upon the convenience and financial interests of the claimants. No doubt it is to their interest to have every detail cut and dried on the day the contract is signed, but the contract does not contemplate that. It contemplates further details and instructions being provided, and the engineer is to have a time to provide them which is reasonable having regard to the point of view of him and his staff and the point of view of the Corporation, as well as the point of view of the contractors.
>
> In determining what is a reasonable time as respects any particular details and instructions, factors which must obviously be borne in mind are such matters as the order in which the engineer has determined the works shall be carried out (as he is entitled to do under clause 2 of the specification), whether requests for particular details or instructions have been made by the contractors, whether the instructions relate to a variation of the contract which the engineer is entitled to make from time to time during the execution of the contract, or whether they relate

> to part of the original works, and also the time, including any extension of time, within which the contractors are contractually bound to complete the works.
>
> In mentioning these matters, I want to make it perfectly clear that they are not intended to be exhaustive, or anything like it. What is a reasonable time is a question of fact having regard to all the circumstances of the case, and the case stated does not disclose sufficient details of the circumstances relating to any particular details or instructions to make it possible for me to indicate what would be all the relevant factors in determining what was a reasonable time within which such details and instructions should have been given. What I have mentioned are merely some examples of factors which may or may not be relevant to any particular details or instructions given which the arbitrator has considered.'

In *A. McAlpine & Son* v. *Transvaal Provincial Administration* (1974), a South African case, a motorway contractor asked the court to define an implied term on the time for supplying information and giving instructions on variations as either:

(a) a time convenient and profitable to himself
(b) a time not causing loss and expense
(c) a time so that the works could be executed efficiently and economically.

The court declined on the grounds that under the contract variations could be ordered at any time, irrespective of the progress of the works, and that drawings and instructions should be given within a reasonable time after the obligation arose.

5.6 *Documents to be kept on site*

Clause 7(5) follows the provision in clause 7(4) of the Fifth edition requiring one copy of the drawings and specification to be kept on site at all reasonable times. The provision is extended, however, to cover documents relating to contractor's design.

5.7 *Permanent works designed by contractor*

Clause 7(6) is a wholly new provision which recognises that the contractor may be required to design parts of the works.

It operates only where the contract expressly provides that there should be some element of contractor's design and only in respect of design for parts of the permanent works.

The provision has two limbs. Firstly, under clause 7(6)(a) the contractor must submit to the engineer for approval such drawings, specifications, calculations and other information as necessary to satisfy the engineer on the suitability and adequacy of the design. Secondly, under clause 7(6)(b) the contractor must submit to the engineer for approval operation and maintenance manuals and completed drawings in sufficient detail to enable the employer to operate, maintain, dismantle, reassemble and adjust the work so designed.

Suitability and adequacy

The phrase 'suitability and adequacy of the design' in clause 7(6)(a) could be taken to imply fitness for purpose of the contractor's design. There may be a conflict here with clause 8(2) which requires the contractor to exercise reasonable skill and care in his design. See the note on that clause in chapter 6.

Manuals and drawings

Clause 7(6)(b) anticipates a problem which is not uncommon where contractor design is involved by stating that no certificate under clause 48 covering any part of the permanent works shall be issued until operation and maintenance manuals and drawings in sufficient detail have been submitted to and approved by the engineer.

A certificate under clause 48 is a certificate of substantial completion and the withholding of this can be seen as a powerful incentive for the contractor to perform under clause 7(6)(b) or a powerful sanction if he fails.

The procedure would seem to be that under clause 48(2) the engineer would specify what was lacking before a certificate would be issued. There may however be a defect in clause 48 as far as parts of the works occupied or used by the employer prior to completion of the whole are concerned. Clause 48(3) simply says if the employer occupies or uses parts of the works the contractor may request in writing and the engineer shall issue a certificate of substantial completion. There is no power for the engineer to refuse a certificate in this case.

5.8 *Contractor's responsibility unaffected by approval*

Clause 7(7) contains two new provisions. Firstly, that approval by the engineer of the contractor's design does not relieve the contractor of his responsibility under the contract. Secondly, that the engineer is responsible for the integration and co-ordination of the contractor's design with the rest of the works.

Approval by the engineer may not relieve the contractor of his responsibilities to the employer but it may well make the engineer liable to the employer in contribution proceedings if there is a failure of the contractor's design. For further reading on this see the case of *Holland Hannen & Cubitts* v. *Welsh Health Technical Services Organisation* (1985).

Similarly the engineer is taking on a burden by accepting responsibility for the integration and co-ordination of the contractor's design and he would be unwise to take this on lightly or without recompense from the employer to whom he is liable if things go wrong.

Chapter 6

General obligations

6.1 *Introduction*

This chapter covers clauses 8 to 19 inclusive dealing with the general obligations of the parties.

The most significant changes have been made in:

clause 11 – provision of site information
clause 14 – contractor's programme.

Changes in detail have been made in:

clause 8 – contractor's responsibilities
clause 12 – unforeseen conditions
clause 13 – compliance with engineer's instructions
clause 19 – safety and security.

Minor drafting changes are to be found in:

clause 9 – contract agreement
clause 10 – performance security
clause 15 – contractor's superintendence
clause 16 – removal of contractor's agent.

In this section only two clauses have remained wholly unchanged between the Fifth and Sixth editions:

clause 17 – setting out
clause 18 – boreholes and exploratory excavation.

6.2 *Clause 8 – contractor's responsibilities*

General responsibilities

Clause 8(1) states the contractor's general responsibilities to construct and complete the works in identical words to clause 8(1) of the Fifth edition with only the word 'maintain' omitted.

The new layout of the clause however separates the responsibilities into sub-clauses:

8(1)(a) – construct and complete
8(1)(b) – provide all labour, materials etc.

Construct and complete

The general responsibility to construct and complete is also found in the form of tender and the form of agreement and, if expressed without any qualification, it would place a strict obligation on the contractor to complete whatever the difficulties – with only 'frustration' in the legal sense available as relief. Basically the employer does not warrant that the works can be built; it is the contractor who makes the offer and undertakes to build.

But the contractor can take some comfort from the opening words of clause 8(1) 'subject to the provisions of the Contract'. There are other provisions in the Conditions, particularly those in clause 13 and clause 51, which require the engineer to issue instructions or variations when certain difficulties arise. Thus the employer does to some extent share the contractor's responsibilities for constructing and completing the works.

Provide all resources

The responsibility in clause 8(1)(b) for the contractor to provide all labour, materials etc. for construction and completion would again be a strict obligation if it were not for the qualifying words 'so far as the necessity for providing the same is specified in or reasonably to be inferred from the Contract'.

The clause has obvious application where the employer has promised to supply some of the resources, usually materials but occasionally plant or labour. But arguments on the clause mostly relate to contractors' claims for payment for resources the contractor is obliged to provide.

A stock response to such claims is to say that clause 8(1)(b), taken with clause 11(3)(b) – (rates and prices to cover all obligations under the contract), obliges the contractor to provide all labour, materials etc. as necessary. To this the contractor can reply with some confidence that clause 8 does not say anything about payment and the proviso in clause 11(3)(b) 'unless otherwise provided in the Contract' opens the door to payment under 30 or more clauses of the contract.

Contractor's design responsibility

Clause 8(2) confirms that the contractor is not responsible for the design of any part of the permanent works except as expressly provided in the contract. A similar provision was tucked into the Fifth edition in clause 8(2) under the side note – 'Contractor's responsibility for the safety of the site operations'.

It is not clear if the words 'except as may be expressly provided in the Contract' mean only that there is to be no implied responsibility on the contractor for design of the permanent works or if they also mean that any such design responsibility is to be expressed at the outset. The difficulty is that 'the Contract' as defined in clause 1(1)(e) includes specifications and drawings which can by clauses 1(1)(f) and 1(1)(g) be modified or added to from time to time. In short, can design responsibility be placed on the contractor by a variation under clause 51? The answer is probably not. The engineer certainly cannot use clause 51 to change design responsibility from himself to the contractor and if he is entitled to order additional work with contractor's design there seems to be a need for practical considerations for some measure of agreement on whether the contractor has the necessary design resources and capability and can undertake the obligation.

For further comment on this see chapter 16 on the use of provisional sums and prime cost items.

Reasonable skill and care

Clause 8(2) provides that the contractor shall use reasonable skill, care and diligence in designing any part of the permanent works.

This is apparently intended to put the contractor in the same position as does the design warranty in the standard building form with contractor's design (JCT 81). This gives the contractor like liability 'as would an architect or, as the case may be, other appropriate professional designer holding himself out as competent to take on work for such design'.

The effect of the JCT 81 warranty is to limit the contractor's design responsibility to one requiring proof of negligence and to exclude the much wider responsibility for fitness for purpose. Whether or not this has been achieved in clause 8(2) of the Sixth edition is debatable.

The clause appears to do no more than confirm responsibility for skill and care. It may have no limiting effect on fitness for purpose. It can certainly be argued that the words 'suitability and adequacy of design' in clause 7(6)(a) imply fitness for purpose.

Contractor's responsibility for site operations

Clause 8(3) repeats exactly the opening words of clause 8(2) in the Fifth edition, namely, that the contractor shall take full responsibility for the adequacy, stability and safety of all site operations and methods of construction.

This is a broad statement which appears to place unqualified responsibility on the contractor. But the impression is misleading.

Firstly there is the position under statute, particularly the Health and Safety at Work Act 1974. The duties under that Act apply to 'any person' and they cannot be excluded by private contract.

Then there are express contractual qualifications. Clause 19 details the responsibilities of the employer for his own workmen and contractors and clauses 20 and 22 list 'excepted risks' and 'exceptions'.

But of more concern are recent court decisions which reveal very clearly that 'full responsibility' under this clause can be severely diminished by other contractual provisions.

The effects of clause 12

In *Humber Oil Terminals Trustee Ltd* v. *Harbour and General Works (Stevin) Ltd* (1991), the Court of Appeal had to consider whether the contractor's obligation under clause 8(2) affected his rights of claim under clause 12 for unforeseen conditions. The contract was for the construction of mooring dolphins and the contractor selected and used a jack-up barge equipped with a fixed crane. As the crane was lifting and skewing a concrete soffit member it became unstable and collapsed. The contractor submitted a claim under clause 12 that he had encountered physical conditions which could not have been foreseen. The employer disputed this claim, of which more will be said later in this chapter, and argued that even if the contractor did encounter such conditions he had no claim under clause 12 by virtue of the provisions of clause 8(2).

The arbitrator, the judge at first instance and the Court of Appeal all held that the contractor succeeded on both point 1 – that he had met unforeseen physical conditions – and point 2 – that his claim was not excluded by clause 8(2). This is how the Court of Appeal reached its decision on point 2:

> 'As to point 2, both the arbitrator and the judge disposed of the point very briefly. The [employer's] argument is a simple one. Clause 8(2), they say, imposes an unqualified full responsibility for the adequacy, stability and safety of all site operations and methods of construction. Here they say the operation or method of construction, to wit by Stevin 73, was inadequate unsafe and unstable. Therefore the consequences are wholly a matter for the contractor and had nothing to do with their employers.
>
> There is in clause 8(2) no exception such as is to be found in clause 8(1) and in other clauses in the contract where expressions such as 'subject to the provisions of the contract' appear. [Counsel for the employer] referred us to a number of such provisions. There is thus, says he, no room for the operation of clause 12. Clause 8(2) is unqualified and the fact that it is so unqualified is reinforced by the presence of qualifications in such other conditions.
>
> Against this it is contended that clause 12 is also unqualified, save as to weather conditions and the like, and it makes no commercial sense, so it is said, to construe the contract as giving no effect to clause 12 in a situation such as the present.
>
> The matter is in my view one largely of first impression. For my part however I am unable to accept a construction in which such inadequacy, unsafety or instability as occurred was due to unforeseen physical conditions intended to be excluded by clause 8(2) from the operation of clause 12. The construction which is put on it by the [employer] involves that, in situations such as this where, as a result of the finding on point 1, the collapse was due to unforeseen physical conditions, clause 12 will have been emasculated to the point of disappearance.'

Lord Justice Nourse, agreeing, said this:

> 'I agree with my Lord that the second question is really one of impression. I do not feel able to say, as [counsel for the contractor] has suggested, that this is a case where there was no inadequacy or instability of site operations or methods of construction within clause 8(2) of the ICE Conditions. On the whole I think that there was. On the other hand, I cannot construe clause 8(2) as applying to a case where the inadequacy or instability is brought about by the contractor's having encountered physical conditions within clause 12(1). That I think was the instinctive view of the arbitrator, who dealt with this question very briefly. Like the judge, I have not found it an easy question, but like him and my Lord I would on balance decide it in favour of the contractors.'

The decision in the *Humber Oil Terminals* case will cause great concern to engineers and employers. It shows that clause 12 cuts across the general principle that the contractor should be free to select his own methods of working but in doing so he must take responsibility for them. It means that the cost to the employer of the contract price will be dependant on how susceptible the contractor's chosen methods of working are to unforeseen conditions. This will not be acceptable to many employers.

As for the engineer, will he be responsible to the employer if he gives consent to contractor's methods under clause 14 which are so susceptible? And will he feel obliged to demand a belt and braces approach to all temporary works?

The effects of clauses 13 and 51

In *Yorkshire Water Authority* v. *Sir Alfred McAlpine & Son (Northern) Ltd* (1985) the contractor was required to submit with his tender for works at Grimwith Reservoir a method statement showing that he would work upstream in constructing an outlet tunnel.

This method statement became listed as one of the contract documents which included the ICE Fifth edition Conditions. In the event the contractor claimed it was impossible to work upstream and worked downstream. He then claimed he was entitled to a variation under clause 51 and payment accordingly.

The employer, relying strongly on clause 8(2), argued that the adoption of a new method statement and a new method of working was entirely the contractor's responsibility. It was said that, if performance according to the method statement was impossible, it must be the responsibility of the contractor to provide an alternative method statement, and even if he was entitled to a variation order, he must, by virtue of the words 'full responsibility', bear any extra cost which this variation might involve.

This argument was rejected by Mr Justice Skinner who said that, where there was a pre-specified method of construction, clause 8 was relevant only to post-contractual methods submitted under clause 14.

More will be said on the *Yorkshire Water* case later in this chapter and in chapters 10 and 13 but it is now clear, if it was not already clear before, that restrictions on the contractor's freedom of choice of methods of working diminish the contractor's responsibilities under clause 8. As Mr Justice Skinner said:

> 'In my judgment, the standard conditions recognise a clear distinction between obligations specified in the contract in detail, which both parties can take into account in agreeing a price, and those which are general and which do not have to be specified pre-contractually.
>
> In this case the applicants could have left the programme and methods as the sole responsibility of the respondents under clause 14(1) and clause 14(3).
>
> The risks inherent in such a programme or method would then have been the respondents' throughout. Instead, they decided they wanted more control over the methods and programme than clause 14 provided. Hence clause 107 of the specification; hence the method statement; hence the incorporation of the method statement into the contract imposing the obligation on the respondents to follow it save in so far as it was legally or physically impossible. It therefore became a specified method of construction by agreement between the parties.'

6.3 Clause 9 – contract agreement

Clause 9 requires the contractor 'if called upon' to enter into and execute a contract agreement in the form annexed to the Conditions. The only change from the Fifth edition is that 'if called upon' is substituted for 'when called upon'. The employer remains responsible for the cost of preparing the contract agreement.

The change from 'when' to 'if' may have been made in recognition of the apparent option in the form of tender which says 'Unless and until a formal Agreement is prepared...'

Without a formal written agreement the legal limitation period for the contract cannot be extended from the standard statutory six years to the 12 years which applies to contracts executed as a deed (previously called contracts under seal).

An interesting legal position would apply if the contractor was called on to enter into a contract agreement and refused to do so – a not unlikely refusal – since not every contractor realises that an extended limitation period can operate in his favour in submitting claims for breach and it is not simply a device for extending his liabilities.

Refusal would clearly be a breach of contract – but would the employer have any effective remedy? An order for specific performance would seem to be the answer, but this is something more likely to be obtained through the courts than through arbitration.

Point of formation of contract

There is little doubt in the minds of most civil engineers that a tender and a letter of acceptance constitute a binding contract. The standard form of tender used with recent editions of the ICE Conditions appears to be quite definite on the point: 'Unless and until a formal Agreement is prepared and executed this Tender together with your written acceptance thereof, shall constitute a binding Contract between us.'

But in the *Yorkshire Water* case referred to in Section 6.2 above, the judge took a different view. He said:

> '[Counsel for the contractor] argues that that contract was formed when the letter of acceptance was posted. But [counsel for the employer] asks, 'A contract to do what?' He argues that it is merely a contract to contract. In my judgment, it is plainly a provisional document and it contemplates no assumption of obligations by either party to perform any of the works until a formal contract has been concluded.'

It may be that the substitution of 'if' for 'when' in clause 9 will be relevant in any future arguments on the point but for those who prefer certainty in the formation of their contracts the solution is to sign the form of agreement at the outset.

6.4. *Clause 10 – performance security*

Clause 10 deals with what are usually called 'performance bonds'.

It states that if the contract requires the contractor to provide security for performance, it shall be provided by a body approved by the employer and shall be in the form of bond annexed to the Conditions.

The appendix to the form of tender shows whether or not a bond is required and if so the amount, expressed as a percentage of the tender total. Clause 10(1) limits this to 10%.

Alternative security

The alternative security of the Fifth edition, 'two good and sufficient sureties' is omitted from the Sixth edition. It was not widely used and it lacked the precision of a formal bond.

However some contractors have only limited bonding facilities and alternatives such as cash deposits, deductions from early certificates and increased retentions are increasingly being offered.

Non-provision of bond

If a bond is required it is to be provided within 28 days of the award of the contract. Unlike many standard forms of building contract, there is no express provision in the ICE Conditions for the employer to determine the contractor's employment if the bond is not provided. However, the breach may be sufficiently serious to justify common law determination and support for this is to be found in a South African case, *Swartz & Son (Pty) Ltd* v. *Wolmaransstadt Town Council* (1960).

Payment for cost

Clause 10 requires the contractor to pay the cost of the bond unless the contract provides otherwise. There will not normally be any dispute on this.

The contractor, however, should be careful in providing a bond as a post-contract requirement without first clarifying who is to pay the cost. In *Perini Corporation* v. *Commonwealth of Australia* (1969), after formal acceptance of the tender, the employer required the contractor to provide a bond. The contractor was unable to recover the cost. This is what the judge said:

> 'The reality of the situation in my opinion is that the plaintiff and the defendant agreed to the provision by the plaintiff of an additional guarantee but that they did not make any agreement at all with respect to the liability of one side or the other side for the cost of doing so. It is my opinion simply a matter upon which the parties have not expressed any agreement and for that reason the claim of the plaintiff on this point must fail.'

There may be a general lesson in this in that assumptions on payment are not always as well founded as they seem.

The ICE model form of bond

Bonds differ considerably in their drafting and the conditions under which they can be called in for payment. At one end of the scale there are 'demand' bonds which can be called in without proof of default; at the other end there are performance or 'conditional' bonds which can only be called upon with certification of default by the contract supervisor.

The ICE model form of bond is a performance or conditional bond and although it does not expressly require a certificate of default, it does rely on both proof of default and proof of damages. This was confirmed in the Hong Kong case of *Tins Industrial Co. Ltd* v. *Kono Insurance Ltd* (1987) where the dispute turned on whether a bond with wording identical to that of the ICE model was absolute (on demand) or conditional on performance. Mr Justice Hunter after commenting on the wording:

> 'Now the trouble about this bond is that it is written in thirteenth century (or earlier) language. It is archaic language and therefore difficult to read and to understand'

went on to re-affirm that it was a performance bond. He said:

> 'One starts with this: the bond is conditional not absolute.

Simply taking the contractor's position, what the contractor is saying is 'if I perform the bond is null and void: and therefore if I perform the bond is discharged'.

If you turn that round it means: 'If I do not perform I have to pay'. You then ask the question who has to say whether you performed or not? You readily come to the conclusion, we suggest, that he who asserts breach has to prove breach; he who claims under the bond has to prove those breaches.'

Release of bond

The model form of bond provides a surety which remains in force until the defects correction certificate. Most contractors will offer, as an alternative, a bond which expires on the issue of the certificate of substantial completion. The reason quite simply is that the longer the bonds remain in force, the more of the contractor's bonding facility is used.

Disputes on bonds

Clause 10(2) supports the provision in the standard form of bond for arbitration on any dispute regarding the date of the defects correction certificate. The resolution of such a dispute takes place outside the scope of clause 66 and is said by clause 10(2)(b) to be without prejudice to any dispute under that clause.

6.5 *Clause 11 – provision of information and sufficiency of tender*

Clause 11 covers two closely related matters – the provision of information on the site by the employer and the obligation of the contractor to inspect and examine the site to ensure the sufficiency of his rates and prices.

The changes to this clause have attracted as much attention and as much concern as any in the Conditions, not least because there is now a clear obligation on the employer where none was expressed before. That is that the employer shall make available to the contractor, for tendering purposes, all the information he has on ground conditions. In support of this it is said that this is no more than a matter of commonsense and good practice; that the employer has everything to gain from full disclosure; and that the contractor's responsibilities are not in any way diminished.

But that approach fails to recognise what will be apparent to anyone comparing the two editions: namely that clause 11 as now presented suggests a radically different approach to the balance of risk. In the Fifth edition the clause starts with an obligation on the contractor to examine and inspect the site; in the Sixth edition the clause starts with an obligation on the employer to make available all information. In the Fifth edition the clause ends that the contractor is deemed to have satisfied himself on the sufficiency of his rates and prices; in the Sixth edition the clause ends with the further proviso that the contractor is deemed to have based his tender on the information made available by the employer.

Even if that is not meant as an invitation to claims it will certainly be taken as one.

Information – general legal position

The general position on responsibility is that, in the absence of specific information and express provisions in the contract to the contrary, the contractor takes all risks from adverse ground conditions. In the old case of *Bottoms* v. *York Corporation* (1892), where neither the contractor nor the employer took boreholes on a sewerage scheme adjacent to a river, it was held that the contractor was not entitled to abandon the contract for difficulties in dealing with water in the excavations since the employer had given no guarantee or representation on the nature of the ground.

In most construction contracts, however, the employer does supply some information and there are usually some express provisions relating to that information and the balance of risk for unforeseen ground conditions. The legal position in such circumstances on the accuracy or adequacy of information given can be exceedingly complex. It can touch on the subject of misrepresentation and it can bring in the provisions of the Misrepresentation Act 1967. That act not only extends liability for misrepresentation beyond fraud – so that innocent representations made without reasonable grounds for belief give potential liability – but it also makes terms in contracts excluding liability of no effect unless they satisfy a test of reasonableness.

Some pointers to the way the law works can be gained from the following cases:

(a) In *Pearson & Sons* v. *Dublin Corporation* (1907) the size of an undersea wall which the contractor was to use was incorrectly marked on drawings by the engineer who had taken no steps to verify the accuracy of his dimensions. It was held that a clause in the contract requiring the contractor to satisfy himself on all dimensions could give protection only against honest mistakes but the engineer's conduct amounted to fraudulant misrepresentation.

(b) In *Morrison-Knudsen* v. *Commonwealth of Australia* (1972) a dispute arose on a civil engineering contract with the requirement that the contractor should inform himself as to site conditions and a disclaimer on site information provided by the employer. The contractor brought an action against the employer for breach of a duty of care by supplying information that was false, inaccurate and misleading. On the facts the court declined to say that the employer had a duty or care or that the contractor had no cause of action but on the matter of the contractor's obligation to inform himself it was said:

> 'The basic information in the site information document appears to have been the result of much highly technical effort on the part of a department of the [employer]. It was information which the [contractors] had neither the time nor the opportunity to obtain for themselves. It might even be doubted whether they could be expected to obtain it by their own efforts as a potential or actual bidder. But it was indispensable information if a judgment were to be formed as to the extent of the work or be done in making the landing strips of the proposed airport.'

(c) In *Bacal Construction Ltd* v. *Northampton Development Corporation* (1975) on a dispute in a design-build contract as to the accuracy of ground condition information supplied by the employer, it was held that there was an implied term that the ground conditions would be as indicated and that the employer was liable for the cost of the redesigned work.

Clause 11(1) – provision and interpretation of information

By clause 11(1), which is completely new, the employer is 'deemed' to have made available to the contractor, before submission of the tender, all information on ground conditions 'obtained by or on behalf of the Employer from investigations undertaken relevant to the Works'.

Lawyers have already warned that the word 'deemed' gives perverse interpretation of clause 11(1) – namely that although the employer may provide nothing at all, whatever he has got is deemed to have been given. But this is obviously not the intention of the clause. The word 'deemed' apparently is only used in this clause because of the legal difficulty of imposing in the contract pre-contract obligations – a situation which would arise if the word was omitted.

Taking the sensible interpretation, it is clearly intended that the employer, usually through his engineer in practice, will have an open record policy with all tenderers. It is not necessary to include all the relevant information with the tender documents but it is advisable to list everything which could be relevant and make it available for inspection.

Employers and engineers should give the widest possible interpretation to what is meant by 'relevant to the Works'. This is a clause where it is infinitely safer to give too much rather than too little.

Liability for damages

The danger for the employer in this new obligation is that, if he is found to be in breach, he may be liable for damages. The contractor may say 'if only I had been given that information my price would have been higher' or perhaps even 'I would not then have tendered at all'. The contractor, of course, may not know until he goes 'fishing' precisely what information has not been disclosed but that will be automatically remedied in any arbitration or litigation proceedings at the discovery stage when both parties must reveal all documents relevant to the matter in dispute.

Interpretation of information

The second provision in clause 11(1) is that the contractor shall be responsible for the interpretation of all information 'for the purposes of constructing the Works' and any design for which he is responsible.

This is a curiously worded provision because the primary purpose of the information made available by the employer under clause 11(1) is not for constructing the works but for the submission of tenders. It is not clear whether the provision is intended to emphasise that the contractor is not responsible for interpretation when tendering, but he is responsible for interpretation when it comes to carrying out his obligations under the contract, or whether it is intended that he should be responsible for interpretation in both situations.

Clause 11(2) – inspection of site

Clause 11(2) corresponds with clause 11(1) of the Fifth edition but with the reference in that clause to information which may have been provided by the employer omitted. The contractor is deemed:

(a) to have inspected and examined the site
(b) to have satisfied himself so far as is practicable and reasonable on:

  (i) the form and nature of the ground
  (ii) the extent and nature of the work
  (iii) the materials necessary for construction
  (iv) the means of communication to the site
  (v) the access to the site
  (vi) the accommodation required

(c) to have obtained all information on risks, contingencies and other circumstances.

'So far as practicable and reasonable'

The proviso 'so far as practicable and reasonable' appears to recognise the difficulties tenderers face in the time available to them and the restrictions they face in carrying out their investigations. However, it is only in respect of ground conditions that the contractor gets express relief by way of payment in appropriate circumstances through clause 12. For other matters, including those listed, the contractor would normally take full responsibility and did so under the Fifth edition when the proviso was attached only to ground conditions.

It is unlikely that the application of the proviso in the Sixth edition to wider matters is intended to indicate risk sharing on those matters but how else can the words be interpreted? The obligation on the contractor has clearly been modified and the change will be of the greatest interest to lawyers and claims consultants.

Clause 11(3) – basis and sufficiency of tender

The wording of clause 11(3) strongly reinforces the proposition that the contractor's rights of claim are greatly improved under the Sixth edition.

Clause 11(2) of the Fifth edition which placed responsibility for sufficiency of rates and prices firmly on the contractor is reproduced as clause 11(3)(b), but the new clause 11(3)(a), taken with the new provisions in clause 11(1) and the modifications in clause 11(2), leaves the employer seriously exposed.

Clause 11(3)(a) says the contractor is deemed to have based his tender on the information made available by the employer and by his own inspection and examination. In the event of conflict between these two it is not clear which is intended to prevail. Whichever is selected knocks a big hole in the 'foreseeability' test in clause 12. It is apparently not for the contractor to foresee anything in his tender – he is deemed to have based it on that which can be seen, either from information provided by the employer or from his own inspection and examination.

Correctness and sufficiency of rates

The reference in clause 11(3)(b) to 'the correctness and sufficiency of the rates and prices stated ... in the Bill of Quantities' is sometimes taken as an indication that each and every rate is to be correct and sufficient in itself – a point of some interest in relation to rate fixing and revaluations. But the reference, which is the same in the Fifth edition, is not expressed as a contractual requirement for such individual self-sufficiency and it is probably intended more as a general statement of composite obligations since few, if any, rates in themselves could be said to cover 'all of his obligations under the Contract'.

6.6 *Clause 12 – adverse physical conditions*

Clause 12 remains essentially the same as in the Fifth edition in providing that the employer rather than the contractor takes the risks of delay and extra cost arising from adverse physical conditions and artificial obstructions which could not have been foreseen by an experienced contractor. But there are three changes to note:

(a) the requirement for the contractor to give notice of encountering such conditions whether or not he intends to claim
(b) the additional action for the engineer to request information on alternative measures
(c) the addition of profit to all the contractor's costs in any claim.

Balance of risk

Civil engineers who have worked solely with the ICE Conditions are often surprised to learn that many standard forms of building contract have no equivalent to clause 12 and the contractor takes the risk of the unforeseen. It is simply a matter of balance of risk; and in civil engineering, where so much of the work is in the ground or underground, it is generally thought to be cheaper for employers to pay for what does happen rather than what might happen. That, of course, assumes that contractors would build allowances into their prices for the unforeseen if they carried the risk. That is a very big assumption.

Effect of clause 12

In *Holland Dredging* v. *Dredging & Construction Ltd* (1987) Lord Justice Purchas in the Court of Appeal likened clause 12 to a shield protecting the contractor against the provisions of clause 11. By that he meant that, without clause 12, the provisions in clause 11 for rates and prices to cover all obligations would prevail in all circumstances, foreseeable or not. Indeed it was just as well for the contractors that the Court of Appeal took the view that it did, for the judge at first instance in the case had held that clause 12 applied only to supervening events – that is, those arising after the contract was formed, not those in existence at the time of tender. The judge had said:

> 'The unqualified deeming effect of clause 11(1) resulted in the dredging contractor agreeing to bear the risk of unforeseen but existing adverse physical conditions and artificial obstructions whether or not an experienced contractor would have reasonably foreseen them.
>
> The words in parenthesis in sub-clause (2) of clause 11 tie in with the general proposition that, where a contractor bears the risk of a site, he must, if he so wishes, include in his price for that risk.... In these circumstances no claim can lie under condition 12.'

If this had not been overturned by the Court of Appeal, the majority of clause 12 claims would be defeated whether under the Fifth or Sixth editions.

Meaning of 'physical condition'

Clause 12 applies when the contractor encounters 'physical conditions (other than weather conditions or conditions due to weather conditions) or artificial obstructions which could not have been foreseen by an experienced contractor'.

Most engineers asked to define the meaning of 'physical conditions' in this context would point to something tangible – boulders in clay; a high water table; running sand; hard rock. Most would take it for granted that clause 12 applies to conditions which are found in the ground. Both views need to be re-assessed following the Court of Appeal decision in *Humber Oil Terminals* v. *Harbour & General Works Ltd* (1991).

In that case, mentioned earlier in section 6.2 because of its impact on clause 8, the question arose whether the contractor had encountered physical conditions within the scope of clause 12. As a 300 tonne crane on a jack-up barge was placing precast soffit units on piles the barge became unstable and collapsed causing extensive damage to the works, plant and equipment. The barge was a total loss and had to be replaced. There was much delay and extra cost.

The contract was under the ICE Fifth edition Conditions and the contractor claimed under clause 12 that the collapse of the barge and its consequences was due to encountering physical conditions which could not have been foreseen by an experienced contractor. The dispute went to arbitration.

The arbitrator gave an award in favour of the contractor finding that, although the soil conditions were foreseeable, clause 12 contains no limitation on the meaning of 'physical condition'; that a combination of strength and stress, although transient, can fall within the term; and that in this case an unforeseeable condition had occurred.

The employer appealed maintaining that the question should be not whether the collapse could have been foreseen, which it clearly could not, but whether physical conditions could reasonably have been foreseen.

The judge at first instance upheld the arbitrator's award but gave leave to appeal.

The arguments advanced for the employer before the Court of Appeal would certainly have found favour with many engineers – namely that a physical condition is something material, such as rock or running sand, and that an applied stress is not a physical condition nor is it something which can be encountered.

The Court of Appeal, however, rejected the submissions and dismissed the appeal.

This is how Lord Justice Nourse dealt with the argument. He said:

> 'I reject these submissions for the following reasons. First, I agree with [counsel for the contractor], that there is nothing to restrict the application of clause 12(1) to intransient, as distinct from transient, physical conditions. Indeed the express reference to weather conditions, albeit by way of exclusion, suggests the contrary. Secondly, while I would agree that an applied stress is not of itself a physical condition, we are not concerned with such a thing in isolation, but with a combination of soil and an applied stress.
>
> Thirdly, and most significantly, as Lord Justice Butler-Sloss pointed out during the course of the argument, it is impossible to speak of a contractor encountering any form of ground, be it rock, running sand, soil or whatever, without recognising that stress of one degree or another will have to be applied, at any rate notionally, to the ground, which will in turn behave, at any rate notionally, in one way or another, no doubt passively in the case of rock, actively in the case of running sand and perhaps unpredictably in the case of soil.
>
> In other words, for the purposes of clause 12(1), you cannot

dissociate the nature of the ground from an actual or notional application of some degree of stress. Without such an application you cannot predict how the ground will behave. In the present case I would say that the condition encountered by the contractors was soil which behaved in an unforeseeable manner under the stress which was applied to it and that that was a physical condition within clause 12(1).'

Logical as this may be it does present problems for the application of clause 12. All collapses on site are unforeseeable – unless there is gross negligence. If a physical condition is something transient such as a combination of stresses then what is there in clause 12 to limit such physical conditions to the ground, about which clause 12 expressly says nothing? Would not a failure in the leg of a crane be as unexpected a physical condition as a failure in the ground beneath the leg?

Clearly the *Humber Oil* case has raised more questions than it has answered and interesting times lie ahead.

Clause 12(1) – notice of conditions

Clause 12(1) now serves only as a requirement that the contractor shall give written notice to the engineer when he encounters physical conditions or artificial obstructions which in his opinion could not have been foreseen.

This is a requirement which clearly applies whether or not the contractor intends to go on to make a claim and its purpose seems to be to put the engineer on notice of potential difficulties.

'Other than weather conditions'

The exclusion for conditions due to weather conditions is taken to apply to immediate weather conditions rather than to the climatic changes which influenced geological strata. But it is still difficult to find a precise boundary. Heavy rain, followed by flood, would clearly create conditions due to weather; but should an unexpectedly high water table in a trench be attributed to weather or a wider cause?

Artificial obstructions

Artificial obstructions are usually taken to be buried, man-made objects and clause 12 can, in appropriate circumstances, serve as an effective claim clause for statutory undertaker's apparatus which is found to be out of position. However, much depends on the accuracy of location indicated by any information supplied. A general statement that a gas main exists in a road suggests that finding the gas main cannot be unforeseen; whereas finding a gas main in the east footpath after being informed it was in the west footpath could be unforeseen. In *C.J. Pearce Ltd* v. *Hereford Corporation* (1968), of which more is said under clause 13 below, the precise location of an old sewer, which the contractor had to cross, was unknown. The sewer collapsed into the contractor's excavation. It was held that, even had the contractor claimed under clause 12, which he did not, his claim would have failed since the condition could have been foreseen.

'Could not reasonably have been foreseen'

There are differences of opinion on whether the foreseeability test is wholly objective or whether it allows for the special knowledge of a particular contractor.

In order to put all tenderers on the same footing it is suggested that the test should be wholly objective and that the phrase 'in his opinion' entitles the tenderer/contractor to take the test as expressed literally. Consider for example the position of a contractor who foresees from his special knowledge that the engineer's design will not work – for example, that piles will be needed for a bridge abutment. Clearly he should not include in his tender for piles because he is entitled to be paid for them as a variation. All he can do is price the tender on the same basis as other tenderers.

Indeed not infrequently the action taken by the engineer under clause 12(4) will be a fair indication of what could, or could not, have been reasonably foreseen. If conditions were not foreseen by the engineer why should the contractor be adjudged to have better foresight?

Clause 12(2) – intention to claim

Clause 12(2) requires that if the contractor intends to claim additional payment or extension of time he shall inform the contractor in writing when giving notice of encountering unforeseen conditions 'or as soon thereafter as may be reasonable'.

A subtle change should be noted here between the Fifth and Sixth editions in the reference to clause 52(4). In the Fifth edition clause 12 simply referred to clause 52(4) which required notice as soon as reasonable. In the Sixth edition clause 12(2) says as soon as reasonable 'pursuant to clause 52(4)' but clause 52(4) itself has a 28 day limit.

It would be prudent for contractors to assume that the 28 day notice limit applies to clause 12 claims as it does to others.

Clause 12(3) – measures being taken

Clause 12(3) isolates and amplifies the provisions, previously in clause 12(1), that when giving notice, or as soon as practicable thereafter, the contractor should give details of:

(a) the anticipated effects
(b) the measure being taken or proposed
(c) estimated costs
(d) anticipated delay
(e) anticipated interference.

The clause does not impose a firm obligation on the contractor to decide what action should be taken. Although such an obligation probably rests on the contractor for methods of construction, it is obviously not for the contractor to decide what should be done where design changes are involved. Consequently the contractor should be wary of taking premature responsibility for clause 12 difficulties.

Clause 12(4) – action by the engineer

By clause 12(4) the engineer has the discretion to respond to any notice given by the contractor under clauses 12(1), 12(2) or 12(3) by:

(a) requiring the contractor to report on alternative methods
(b) consenting to the measures notified
(c) giving instructions
(d) ordering a suspension
(e) ordering a variation.

The engineer's action at this stage is not in itself an admission of employer's liability, except in respect of variations, but there is a practical difficulty for the engineer in holding this position. Whichever specified option is chosen, it seems to imply acceptance of liability. The engineer, therefore, may rightly be reluctant to act until he has formed a view on liability but, as will be seen below, that decision should not be delayed.

The option of giving instructions is additional to that of ordering a variation and it may be that one is intended to apply to temporary works and the other to permanent works. But obviously the engineer should be very careful in giving instructions for temporary works, whether liability under the clause is accepted or not.

The ordering of a suspension under clause 40 does not necessarily entitle the contractor to recover his costs – the suspension may be necessary for the proper execution or for the safety of the works. This is, therefore, one option which may provide the engineer with a measure of breathing space whilst he appraises the situation.

Clause 12.5 – conditions reasonably foreseeable

Under clause 12.5 the engineer is required to inform the contractor in writing as soon as he reaches any decision that the conditions could have been foreseen by an experienced contractor. Note, however, that no time limit is set on reaching a decision.

Engineers should beware of falling into the trap of doing nothing in response to a clause 12 claim – although it has to be admitted that such inactivity is far from uncommon. The danger is that the contractor, having given notice of a claim and the measures he is taking, is entitled to know promptly by clause 12(5) whether or not he is going to be paid and failure by the engineer to respond promptly is a breach which leaves the employer exposed to a claim for damages.

The specific nature of the statement in clause 12(5) that the value of any variation ordered 'previously' shall be included in the contract price seems to imply that, if the engineer has taken any other action under clause 12(4) before deciding that the contractor has no right of claim, the contractor cannot recover any 'previous' costs related to such action. However, it is not thought to mean this since there will clearly be cases where the contractor has contractual rights to payment. It probably means only that there can be no going back on the cost of a variation whatever the engineer's findings.

Clause 12(6) – delay and extra cost

The contractor's entitlement to payment under clause 12(6) of the Sixth edition is improved. Under the Fifth edition the contractor received cost plus profit for additional work done and plant used; and cost for delay and disruption. Under the Sixth edition the contractor is entitled to profit on all costs.

This may be the inadvertent result of drafting change or it may be recognition that clause 12 is not to be likened to other claim clauses which have affinity to breach.

6.7 Clause 13 – work to the satisfaction of the engineer

Clause 13 is little changed from the Fifth edition. Profit is added to cost for any additional work in any claim and it is now expressly stated that delay and extra cost resulting from the contractor's default is not recoverable.

Claims under clause 13

The clause remains something of an enigma. Clause 13(1) which requires the contractor to construct and complete the works 'save insofar as is legally or physically impossible' and to comply with and adhere strictly to the engineer's instructions was originally the whole of clause 13 in the Fourth edition.

The clause still raises interesting questions on what is meant by 'impossible' and how wide is the power of the engineer to give instructions. Clause 13(2) which requires the materials, manner, mode and speed of construction to be acceptable to the engineer, appeared first in the Fifth edition as an adaption of certain provisions from clause 46 of the Fourth. Hence the odd reference in clause 13(2) to the 'speed of construction'. Clause 13(3), which entitles the contractor to claim for instructions disrupting his arrangements so as to cause him to incur cost beyond that foreseeable at the time of tender, was new to Fifth edition. Its introduction was greeted with alarm as providing near limitless opportunities for contractors to raise spurious claims.

In the event there is nothing to suggest that clause 13(3) revolutionalised the profits of contractors as predicted and there is very little evidence that clause 13(3) of itself is a major source of claims. It is, however, widely used to support claims for instructions given under other clauses but there is some doubt as to whether this is what is intended having regard to the specific reference to clause 5 and clause 13(1) in the opening words of clause 13(3). It all depends on whether 'the Engineer's instructions' in clause 13(1) are confined to dealing with the 'legally or physically impossible' or whether they are 'on any matter'. The arguments for each case are well balanced.

Application of clause 13(1)

Surprisingly, it is clause 13(1), old as it is, which continues to provide more interest to lawyers and claims practitioners than clause 13(3). In *C.J. Pearce Ltd* v. *Hereford Corporation* (1968) a dispute arose on an ICE Fourth edition contract. The contractor, having got into difficulties when crossing an old sewer, claimed that he had been given instructions by the engineer on how to deal with the matter and since he was obliged to comply with such instructions he was entitled to payment. The judge declined to accept that the engineer's advice amounted to an instruction and held that in any event there was no financial liability since the contractor was bound to complete the contract works.

A similar decision in principle had been reached in the case of *Neodox Ltd* v. *Swinton and Pendlebury Borough Council* (1958) where Mr Justice Diplock had said:

> 'In a contract in which there is no specific method of carrying out particular operations necessary to complete the works set out, and which provides merely that they shall be carried out under the engineer's directions and in the best manner to his satisfaction, I find great difficulty in seeing how a direction by the engineer intimating the manner in which the operations

> must be carried out in order to satisfy him can be a 'variation of or addition to the works'. It seems to me to be no more than what the contract itself calls for, provided only that the engineer is fair and impartial in making his decision to give such direction.'

The other side of this position – that if the works are impossible to perform as specified the contractor is entitled to a variation – was revealed in *Yorkshire Water Authority* v. *Sir Alfred McAlpine & Son (Northern) Ltd* (1985). In that case under the ICE Fifth edition it was held that the incorporation of a method statement into the contract imposed an obligation on the contractor to follow it save insofar as it was legally or physically impossible. Mr Justice Skinner said:

> 'Here there was a specified sequence or method of construction. If the variation which took place was necessary for the completion of the works because of impossibility within clause 13(1), then, in my judgment, the respondents were entitled to a variation order with the consequent entitlement to payment of the value of such variation as is provided in clause 51(2) and clause 52.'

Impossibility

In the *Yorkshire Water* case it was left for the arbitrator to decide on the issue of impossibility so it is of no help in deciding what is meant in clause 13 by 'impossible'.

Some guidance can be found from the case of *Turriff Ltd* v. *Welsh National Water Authority* (1980) where it was held that the tolerances for precast concrete sewer segments, although not absolutely impossible to achieve, were impossible in an ordinary commercial sense.

The issue of impossibility also arose in the *Holland Dredging* case and there Lord Justice Purchas made a significant point when saying:

> '....when taken in conjunction with clause 64 (frustration), clause 13(1) must relate to something short of a frustrating event, where supervening events are concerned.'

Frustration, which is discussed in chapter 19, can only apply to supervening events as indicated by Lord Justice Lloyd in *McAlpine Humberoak Ltd* v. *McDermott International* (1992) when saying:

> 'If we were to uphold the judge's findings of frustration, this would be the first contract to have been frustrated by reason of matters which had not only occurred before the contract was signed and were not only well known to the parties, but had also been provided for in the contract itself.'

But whether 'impossibility' in the context of clause 13(1) can apply to both supervening events and events existing at the time of the contract is another matter. For practical reasons there would seem to be little doubt that it should apply to both but it may be worth noting that the judge at first instance in the *Holland Dredging* case said this:

> 'The words in parenthesis in sub-clause (2) of clause 11 tie in with the general proposition that, where a contractor bears the risk of a site, he must, if he so wishes, include in his price for that risk....
>
> In these circumstances no claim can lie under condition 12. As regards condition 13 the same consideration applies so far as 'physically impossible' is concerned as in the case of condition 12. Only a supervening event resulting in physical impossibility would enable a contractor successfully to advance a claim.'

Now while the Court of Appeal clearly overruled the judge in respect of clause 12 it is not clear that they did so in respect of clause 13 and it will be interesting to see how this point is treated in future cases.

Clause 13(1) – instructions

Although, as discussed above, the intention and interpretation of clause 13(1) may be a matter for debate it is suggested that the contractor should generally take the view that he is obliged to comply with all the engineer's lawful instructions connected with construction of the works. The contractor has his remedy for unforeseen costs under clause 13(3).

Note however, the point emphasised in the final sentence of the clause that the contractor shall only take instructions from the engineer or '(subject to the limitations referred to in clause 2) from the Engineer's Representative'. This proviso is oddly worded in relation to clause 2 but it presumably means the engineer's representative can act either under delegated powers or under his express powers elsewhere in the Conditions.

Clause 13(2) – mode and manner of construction

Clause 13(2) may be intended to deal with approvals rather than instructions. In which case it has no financial implications unless the engineer causes delay by failing to approve (or disapprove) within a reasonable time.
However it can be interpreted as giving the engineer power to redefine what has been specified in favour of what is acceptable. Any such change would almost certainly amount to a variation under clause 51.

Clause 13(3) – delay and extra cost

The provisions and procedures of clause 13(3) can be scheduled as follows:

- if the engineer issues directions
- which involve the contractor in delay or disruption
- causing him to incur cost beyond that foreseeable by an experienced contractor at the time of tender
- the engineer is to determine any extension of time to which the contractor is entitled
- the contractor shall be paid the amount of such cost as is reasonable
- except to the extent that such delay or extra cost results from the contractor's default
- profit shall be added to cost in respect of additional permanent or temporary work
- an instruction requiring a variation is deemed to have been given pursuant to clause 51.

6.8 *Clause 14 – contractor's programme and methods of construction*

Clause 14 of the Sixth edition continues the progression from the

Fourth to the Fifth editions in expanding the detail of the provisions for submission and approval of the contractor's programme and methods and in expanding the extent of the engineer's involvement. The changes are numerous and significant and need to be scrutinised with care by engineers and contractors.

For summary purposes the more important changes can be listed as:

(a) The contractor is expressly required to take account in his programme of possession restrictions under clause 42.
(b) The contractor is required to provide details of his arrangements and methods at the same time as submission of his programme and without any request from the engineer.
(c) The engineer is required to accept, reject or request further information on any programme within 21 days.
(d) The contractor is required to submit a revised programme within 21 days of any rejection.
(e) If the engineer fails to act within 21 days the programme as submitted is deemed to be accepted.
(f) Requests for further information or a revision of programme to ensure completion on time are subject to the same time limits and procedures as above.
(g) The engineer has to state within 21 days whether he consents or rejects the contractor's methods of working whether or not he has requested such information.
(h) Profit is added to cost for additional or temporary work in claims.

Status of clause 14 programmes

Nothing has changed between the Fifth and Sixth editions to alter the contractual status of a clause 14 programme and it is not a contract document. It only comes into existence after the contract has been made. At first sight it may seem odd that it is a breach of contract not to submit a clause 14 programme but failure to follow the programme is not itself a breach. But as shown elsewhere in this book in comment on the *Yorkshire Water* case, whereas programmes and method statements which are fixed pre-contract and are bound into the contract are binding on both parties, programmes and method statements submitted post-contract do not alter the obligations of either party.

As an exception to that general rule, however, it is, of course, necessary to note any specific references in the contract to compliance with any post-contractual submission. Thus under both the Fifth and Sixth editions, the employer is required under clause 42 to give possession of the site in accordance with the clause 14 programme – although that is qualified to the 'accepted' programme under the Sixth edition.

Shortened programmes

Contractors have for many years believed that by submitting programmes showing completion in a shorter time than that allowed in the contract they improve their claims prospects.

It is certainly true that the shorter the time for completion, the more pressure there is on the employer and the engineer in undertaking their obligations and duties, and the greater the likelihood that acts of prevention will delay the contractor. But the question which has to be asked is this: does the entitlement which undoubtedly exists in most standard forms, and certainly in the Fifth and Sixth editions, for the contractor to finish early also entitle him to impose obligations on the employer in respect of achieving an early finish?

The answer to this is clearly 'No', but surprisingly it was not until the building case of *Glenlion Construction Ltd* v. *The Guinness Trust* (1987) that a definitive judgment on the point was given. In that case, the contract allowed 114 weeks for completion and Glenlion, the contractor, submitted a programme showing completion in 101 weeks. One of the questions the judge had to answer was put as follows:

> '...Whether there was an implied term of the contract between the applicant and the respondent that, if and insofar as the programme showed a completion date before the date for completion the employer by himself, his servants or agents should so perform the said agreement as to enable the contractor to carry out the works in accordance with the programme and to complete the works on the said completion date.'

The judge said:

> 'The answer to the question must be "No". It is not suggested by Glenlion that they were both entitled and obliged to finish by the earlier completion date. If there is such an implied term it

imposed an obligation on the Trust but none on Glenlion.
It is not immediately apparent why it is reasonable or equitable that a unilateral absolute obligation should be placed on an employer.'

Effect of approvals

Although clause 14(9) of the Sixth edition and clause 14(7) of the Fifth state that acceptance of the contractor's programme or consent to his methods does not relieve the contractor of any of his duties or responsibilities under the contract, that is not to say that such acceptance or consent cannot impose obligations on the employer or the engineer.

For the employer there is the matter mentioned above in respect of possession of the site in that he is bound by any approval given by the engineer; and there are the provisions in clause 14 for recovery by the contractor for delay and cost caused by the engineer's consent being unreasonably delayed.

For the engineer there is the possibility of liability to the employer in a claim for contribution under the Civil Liability (Contribution) Act 1978 if he negligently approves or fails to disapprove; and there is the possibility of claims from third parties in negligence actions arising out of accidents – see comment later in this chapter under clause 19 (safety and security).

Clause 14(1) – programme to be furnished

Under clause 14(1) (a) the contractor is required to submit within 21 days of the award of the contract a programme showing the order in which he proposes to carry out the works. In the Fifth edition, the requirement was for submission within 21 days of acceptance of tender and the programme had to show the 'order of procedure'. Neither of these minor changes is likely to be of significance.

What causes a little more concern is that it is possible, although certainly perverse, to read clause 14(1) (a) as being no more than a requirement for a programme which complies with the possession of site restrictions prescribed in the contract. But clearly the final phrase in the clause 'having regard to the provisions of clause 42(1)' is really intended to indicate something to be taken into account in the programme, not the purpose of it.

Order of procedure

The old argument on whether clause 14 requires a programme with both a timescale and an order of procedure has not been directly resolved. There is more in the Sixth edition than in the Fifth to imply a timescale and the new powers of the engineer under clause 14(2) can deal swiftly with any omission.

Nevertheless many engineers will continue to phrase their consents as approving only 'order' and saying nothing on timescale. Unless the engineer has good reason to concern himself with the approval of timescale this is a wise policy.

Where the contractor does include timescale in his programme he is, of course, entitled to programme to complete early.

Failure to submit a programme

As with the Fifth edition there are no express sanctions on the contractor if he fails to comply with the submission requirements of clause 14.

Two contractual remedies are available to the employer but both are extreme; the engineer could order a suspension under clause 40 and refuse the contractor's costs on the grounds the programme was 'necessary for the proper execution or for the safety of the Works'; or the engineer could activate the determination procedures of clause 63 on the grounds the contractor was persistently in breach of his obligations under the contract.

The employer has the ordinary common law remedy of damages for breach but it would require unusual circumstances for the employer to be able to prove loss arising from the contractor's default of non-submission of a programme.

General description and methods of construction

Clause 14(1) (b) requires the contractor to provide in writing, at the same time as he submits his programme, a general description of arrangements and methods of construction.

In the Fifth edition the similar requirement for the description and methods appeared to follow a request by the engineer for further details and the phrase 'at the same time' was generally taken to apply to the time of the further details and not the submission of the programme.

The point is now beyond doubt in the Sixth edition but it does have potential claim implications if the engineer delays in consenting to the contractor's methods – see clause 14(8) below.

Rejection by the engineer

Clause 14(1)(c) requires the contractor to resubmit a revised programme within 21 days of any rejection by the engineer. This fills what was sometimes described as a black hole in the Fifth edition which had nothing to say on the procedures to follow rejection.

Note that clause 14(1)(c) applies only to the programme and not to arrangements and methods of construction. Clause 14(7) applies to those matters.

Clause 14(2) – action by the engineer

Clause 14(2) is a much needed new provision requiring the engineer to respond within 21 days of receipt of the contractor's programme by either:

(a) accepting in writing
(b) rejecting in writing with reasons
(c) requesting further information.

If the engineer does none of the above, the programme is deemed to have been accepted as submitted.

In requesting further information on the programme the engineer may be seeking:

(a) clarification
(b) substantiation
(c) satisfaction on reasonableness having regard to the contractor's obligations.

Clearly the engineer has wide scope for making a request for further information but whether it goes so far as entitling the engineer to demand a fully resourced programme, or one with information requirement dates, is another matter.

Clause 14(3) – provision of further information

Clause 14(3) requires the contractor to respond within 21 days to the engineer's request for further information and the engineer in turn to respond within a further 21 days.

Clause 14(3) also provides that, if the contractor fails to provide further information within the 21 day timescale or such other period as allowed by the engineer, the programme shall be deemed to be rejected.

Clause 14(4) – revision of programme

Clause 14(4) corresponds with clause 14(2) of the Fifth edition in entitling the engineer to seek at any time a revised programme showing modifications to the original programme necessary to ensure completion on time.

The clause does not, it is suggested, give the engineer any right to call for a revised programme simply because the contractor is not proceeding in accordance with the original programme. Although the clause uses the phrase 'does not conform' it imposes an obligation on the contractor only in respect of completing on time.

In any event the engineer would be most unwise to call for a revision of programme under this clause without having first reviewed and granted all extensions of time due to the contractor. The subject of constructive acceleration is discussed in chapter 10 under clause 46 but what it amounts to is that, if the engineer uses contractual provisions to force the contractor to complete in a lesser time than that to which he is entitled, the contractor may be able to recover his costs from the employer.

The contractor for his part might consider it appropriate when submitting a revised programme before all outstanding applications for extensions have been dealt with, or where there is dispute on the amount of extension granted, to submit an accompanying statement in the form of preliminary notice of an acceleration claim.

Clause 14(4) links back to the 21 day cycle of times for approvals and resubmissions in clauses 14(2) and 14(3).

Clause 14(5) – design criteria

Clause 14(5) in the Sixth edition is the same as that in the Fifth. It imposes an obligation on the engineer, not on the contractor, to supply design information. To that extent it might have been better placed elsewhere in the Conditions than in a clause which otherwise deals solely with the contractor's obligation to supply information.

The information to be supplied by the engineer is the design criteria necessary to enable the contractor to comply with his obligations for submission of methods of construction.

The provision in clause 14(8)(b) entitling the contractor to claim if he incurs unavoidable cost because of limitations imposed by design criteria supplied under clause 14(5) should be noted carefully by all engineers. Usually such design information will be available at the time of tender and that is when it should be given to the contractor to minimise claims.

Clause 14(6) – methods of construction

Clause 14(6) gives the engineer a discretionary power to request information on methods of construction. The wording of the clause makes clear that it is concerned with detail likely to be beyond that submitted by the contractor originally under clause 14(1).

The purpose of the provision is primarily to enable the engineer to assess the effects of the contractor's methods on the permanent works as built but there is some scope for wider enquiry in the phrase 'constructed and completed in accordance with the Contract'.

It is important for the engineer to note that by using this provision he is burdening himself with the obligation of consenting or rejecting and he is opening the door to claims under clause 14(8) if he delays in his response.

Clause 14(7) – engineer's consent

Clause 14(7) requires the engineer to respond to receipt of any information on the contractor's methods, whether supplied under clause 14(1)(b) or 14(6), by informing the contractor in writing either:

(a) the methods have the engineer's consent, or
(b) in what respect they are deficient.

In the Fifth edition, the engineer was under no obligation to respond to information supplied on the contractor's methods unless he had requested that information.

Clause 14(7) also provides that, where the engineer has stated that the contractor's methods 'fail to meet the requirements of the Contract or will be detrimental to the Permanent Works', the contractor will make such changes as necessary to meet 'the Engineer's requirements' and to obtain his consent.

The change in wording from 'the requirements of the Contract' to 'the Engineer's requirements' is worth noting since it is the phrase 'the Engineer's requirements' which is used in clause 14(8) for claims. The engineer is best advised to remain as far as possible with the 'requirements of the Contract' in anything he has to say. The final provision in clause 14(7) is that the contractor shall not change methods which have received the engineer's consent without further consent in writing. Such further consent shall not be unreasonably withheld.

Clause 14(8) – delay and cost

Clause 14(8) operates when the contractor suffers unavoidable delay or cost because:

(a) the engineer's consent to methods of construction is unreasonably delayed, or
(b) the engineer's requirements on methods or limitations imposed by design criteria could not have been foreseen at the time of tender.

The contractor is entitled to an extension of time and the payment of such cost 'as the Engineer considers fair in all the circumstances'. This is perhaps the vaguest description of the amount recoverable by the contractor to be found anywhere in the Conditions.

Clause 14(9) – responsibility unaffected by consent

Clause 14(9) prevents any possibility of a term being implied that the engineer's acceptance or consent acts to relieve the contractor of his obligations or responsibilities under the contract.

6.9 *Clause 15 – contractor's superintendence*

Clause 15(1) requires the contractor to provide superintendence during the construction and completion of the works, and as long thereafter as the engineer considers necessary.

The phrase 'construction and completion' replaces 'execution' in the Fifth edition and this is the only change made in clause 15.

Contractor's agent

Clause 15(2) requires the contractor to keep a competent and authorised agent constantly on the works giving his whole time to superintendence of the same. The contractual role of the agent is:

(a) to superintend
(b) to be in full charge
(c) to receive instructions on behalf of the contractor
(d) to be responsible for the safety of all operations.

Although contractors sometimes find the burden of keeping a full time agent on smaller contracts uneconomic and onerous, particularly where much of the work has been sub-contracted, it is both dangerous and unwise for engineers to tolerate default. This is a clause which does call for action under clause 40 (suspension) and clause 63 (determination) if there is serious or persistent default.

6.10 *Clause 16 – removal of contractor's employees*

Clause 16 places an obligation on the contractor to employ only careful, skilled and experienced persons and entitles the engineer to require the removal of any person who is incompetent or negligent, guilty of misconduct or incautious of health and safety matters. The only change from the Fifth edition is that provisions on safety in 'the Contract' replace 'in the Specification'.

Taken literally, the requirement of only employing experienced persons does nothing for training or for apprentices but fortunately the criteria for requiring removal could not fairly be applied to diligent trainees or apprentices.

In practice it is unusual for engineers to call for removal of any persons other than on health and safety grounds, with evidence of lunchtime drinking the commonest cause of complaint.

6.11 Clause 17 – setting-out

Clause 17 has three main provisions, shown in the Sixth edition as sub-clauses but otherwise in identical wording to the Fifth edition.

Clause 17(1) confirms that the contractor is responsible for setting-out the works. Clause 17(2) requires the contractor to rectify any errors at his own cost unless the errors arose from incorrect data supplied by the engineer or the engineer's representative. Clause 17(3) confirms that the checking of setting-out by the engineer or the engineer's representative does not relieve the contractor of his responsibility.

Errors discovered after completion

Clause 17(2) retains the phrase 'at any time during the progress of the Works' which leaves open the matter of setting-out errors not discovered until after the works have been completed.

However, the employer can take comfort from the fact that even the issue of the defects correction certificate under clause 61 does not relieve the contractor from his obligations under the contract and he is still liable for damages when they can be proved.

Disputes on setting-out errors

Disputes on setting-out errors commonly arise owing to discrepancies in information from drawing to drawing and the question then becomes to what extent is the contractor obliged to check for consistency? There is no specific obligation on the contractor to do this but it is implied by clause 5 that, if he notices ambiguities or discrepancies, he will refer the matter to the engineer. Although clause 17(2) is without qualification in putting the cost of setting-out errors resulting from incorrect data on the employer, it is suggested that this has to be interpreted with regard to the wider obligations of the parties and the contractor would not be able to recover his costs where he had knowingly proceeded to set out and construct with incorrect data.

The provision in the final sentence of clause 17(3) that the contractor shall carefully protect and preserve all bench marks, sight rails and the like is one that engineers would, no doubt, like to see backed up by express sanctions for default. The work of the resident engineers and site staff is often compounded by lack of protection. But whilst the contract does not provide the employer with express rights of set-off in respect of reckless damage to setting-out material by the contractor, the employer has common law rights if he has to bear the engineer's extra costs.

6.12 *Clause 18 – boreholes and exploratory excavation*

Clause 18 remains exactly as it was in the Fifth edition, a straightforward provision with an unusual capacity for dispute.

The clause provides that if the engineer 'shall require' the contractor to make boreholes or to carry out exploratory excavation such requirement shall be in writing and shall be deemed a variation unless a provisional sum or prime cost item is included in the bill of quantities.

The intention is fairly clear – that the employer will pay for boreholes etc. taken during the progress of the works to enable the engineer to finalise his design. The problem is that the contractor has general obligations to take boreholes etc. to avoid damage to buried apparatus and there is frequently no clear distinction between what the contractor does on his account and what is required for the engineer's purposes. The contractor may be able to argue with some force that the information from his own boreholes is being used by the engineer free of charge. In such circumstances it is suggested that even if the engineer is unwilling 'to require' the contractor to make boreholes, the contractor may be able to include the cost in any variation which results from his discoveries.

6.13 *Clause 19 – safety and security*

Clause 19 on safety and security remains as it was in the Fifth edition, save an added reference to the engineer's representative. Clause 19(1) requires the contractor to have full regard for the safety of all persons entitled to be on the site and to maintain the site in a safe and secure condition. Clause 19(2) places a similar obligation on the employer in respect of any work he carries out on the site with his workforce.

Perhaps because of added emphasis on health and safety matters since 1973 and the introduction of the Fifth edition, greater change was expected in this clause and there has been some criticism that it does not go far enough in spelling out the responsibilities of all concerned in the construction process to have regard to safety. To this the draftsmen have replied that the contractor who is in actual charge of the works should carry the prime responsibility as between the parties to the contract and that it would be inappropriate and misleading to attempt to cover criminal responsibility as imposed by the Health and Safety at Work Act and other statutes. It has been suggested that if special safety provisions are needed for a specific project they should be included in the specification, which itself is a contract document.

Liability of supervisors

The liability of designers for negligence is beyond the scope of this book but the sad case of the Abbeystead pumping station deaths in 1984 and the subsequent case of *Eckersley* v. *Binnie & Partners* (1988) will be on the minds of engineers for many years to come. Two older cases, however, concerning the liability of supervisors in respect of personal injury accidents are worth noting in brief detail.

In *Clayton* v. *Woodman* (1962) a bricklayer was injured when a wall in which he was cutting a chase collapsed on him. The bricklayer sued, amongst others, the architect who had rejected his advice that it would have been better to pull the wall down. The Court of Appeal overturned the trial judge's finding that the architect was liable. Lord Justice Pearson said:

> 'Now it is quite plain, in my view, both as a general propostion and under the particular contract in this case, that the builder, as employer of the workman, has the responsibility at common law to provide a safe system of work, and he also has imposed on him under the Building Regulations the responsibility of seeing that those regulations are complied with, so that everything is as safe for the workman as it reasonably can be. That is the responsibility of the builder, and it is important that that responsibility should not be overlaid or confused by any doubt as to where the builder's province begins or some other person's province ends in that respect. The architect, on the other hand, is engaged as the agent of the owner of the building for whom the

> building is being erected, and his function is to make sure that in the end, when the work has been completed, the owner will have a building properly constructed in accordance with the contract and plans and specification and drawings and any supplementary instructions which the architect may have given. The architect does not undertake (as I understand the position) to advise the builder as to what safety precautions should be taken or, in particular, as to how he should carry out his building operations. It is the function and the right of the builder to carry out his own building operations as he thinks fit, and of course, in doing so to comply with his obligations to the workman.'

In *Clay* v. *Crump* (1963) a wall left standing after a demolition contractor had cleared a site collapsed on a builder's site hut killing two workmen and injuring another. The architect had accepted advice from the demolition contractor that the wall was safe to leave standing without examining it himself. The architect, the demolition contractor and the building contractor were all found liable by the Court of Appeal; the liability of the architect arising from his breach of duty to ensure the site was safe for the main contractor to enter after the demolition work.

Chapter 7

Care of the works and insurances

7.1 *Introduction*

This chapter examines the following clauses:

clause 20 – care of the works
clause 21 – insurance of the works
clause 22 – damage to persons and property
clause 23 – third party insurance
clause 24 – accidents to workpeople
clause 25 – evidence and terms of insurance.

All of the these clauses, with the exception of clause 24, have undergone some change from the Fifth edition. Most of the changes are points of clarification or improvements to procedure but so complex is the subject of insurance that it would be unwise to say categorically that no changes of significance have been made. That will only be determined by useage.

Ordinary users of the Conditions can do no more than recognise that whilst the principles of each clause are straightforward, the interpretation and application of the wording, which is far from straightforward, calls for the assistance of insurance specialists or lawyers when there is any problem with some of degree of difficulty to resolve.

7.2 *Clause 20 – care of the works*

Clause 20 sets out the contractor's responsibility for care of the works. In broad terms that responsibility is already established by the contractor's general obligation as expressed in clause 8 to construct and complete the works and it can, in any event, be established from ordinary legal principles as the contractor's risk.

Purpose of clause 20

The purpose of clause 20 however, is not merely to state the obvious. Its importance lies in:

(a) fixing the time limits during which the contractor carries the risk of care of the works
(b) stating those risks which are excepted from the contractor's responsibility
(c) providing how the costs of rectification of loss or damage are to be borne.

7.3 *Clause 20 – changes from the Fifth edition*

Extent of contractor's responsibility

In clause 20(1) the contractor's responsibility is extended from 'care of the Works' to 'care of the Works and materials plant and equipment for incorporation therein'.

Commencement of responsibility

Also in clause 20(1) the contractor's responsibility now commences on the 'Works Commencement Date' – a date defined by reference to clause 41. Previously a less definitive phrase was used – 'the date of the commencement'.

Completion of responsibility

Again in clause 20(1) the contractor's responsibility now ends on 'the date of issue of a Certificate of Substantial Completion'. In the Fifth edition it ended '14 days after the Engineer shall have issued a Certificate of Completion'.

This is an important change which engineers must note in advising employers to arrange their own insurance. The very useful period of 14 days' grace has been lost. Fortunately it is the 'date of issue' of the certificate and not the date of completion on the certificate which applies, so problems of retrospective liability falling on the employer should be avoidable.

Excepted risks

The Fifth edition defined the 'excepted risks' with a list of events in clause 20(3), albeit in a single overlength sentence of 138 words.

The Sixth edition lists the same events in clause 20(2) but in a more readable fashion and with a sensible adjustment of order, so that those of most practical importance now head the list. There is, however, a serious question to be asked whether the draftsmen have unintentionally, or perhaps deliberately, changed the basis of the definition from the Fifth to the Sixth editions.

The 'excepted risks' in the Sixth edition are not the events themselves as listed in the Fifth edition but are loss and damage to the extent that it is due to those events. The problem phrase is 'to the extent'. This raises the possibility of complex issues of causation but it does suggest that an excepted risk can no longer be apportioned with regard to liability since by definition that apportionment has already taken place.

Rectification of loss or damage

Clause 20(3) of the Sixth edition has substantial drafting changes from clause 20(2) of the Fifth.

The reference in the Fifth to 'injury' which was difficult to understand has been dropped. So has the phrase 'from any cause whatsoever' although if the reason is the old case of *Farr* v. *The Admiralty* (1953) it has taken a long time to come through. In that case it was held, in a contract for building a jetty, that a provision that the contractor should be responsible for making good loss or damage from 'any cause whatsoever' included damage caused when a ship owned by the employer struck the work under construction.

The requirement that at completion the works shall be in 'good order and condition' has also been dropped and all that remains is that they should 'conform in every respect with the provisions of the Contract'.

A useful change for contractors to note is that for any loss or damage arising from the excepted risks the engineer must now give instructions on the extent of any remedial works. The Fifth edition vaguely said 'shall if required by the Engineer'. The Sixth edition says 'shall if and to the extent required by the Engineer'. To achieve this the engineer will have to give instructions and these may amount to variations under clause 51 or instructions under clause 13. This will go someway to assisting the contractor in assessing what he can recover 'at the expense of the Employer' as allowed by clause 20(3) (b).

Apportionment of cost of rectification

Clause 20(3) (c) is a new provision requiring the engineer to carry out an apportionment of cost of rectification of loss or damage which is in part caused by an excepted risk and is in part contractor's risk.

The provision is new in that it expressly requires the engineer to carry out the apportionment. The Fifth edition may have contemplated a similar apportionment but the wording was not as precise. In that edition the phrase in clause 20(2) 'To the extent' could be taken simply as the introduction to the provision for dealing with rectification of excepted risk damage.

The difficulty for the engineer in undertaking the task of apportionment, which the Sixth edition clearly requires that he should, is that although stated in clause 20(3) (c) to be apportionment of cost it also involves apportionment of responsibility. That would seem to be the task for the loss adjusters of the insurance companies involved rather than the engineer.

7.4 Clause 20 – general comment

Completion and useage of sections and parts

Under clause 20(1) (b) the contractor ceases to be responsible for any section or part of the works which has been given a certificate of substantial completion.

This is something the engineer should consider before granting a discretionary certificate under clause 48(4) for any part of the works completed but not put into use. It should also be a factor for the employer to consider before putting any part of the works into premature use; for such use entitles the contractor to a certificate of substantial completion under clause 48(3) if the part is 'substantial'.

But whether or not a certificate is issued, clause 20(2) (a) includes damage due to use or occupation by the employer as an excepted risk. This, however, is not quite the same as a formal transfer of responsibility for a section or part because the test is 'due to' use or occupation and not simply use or occupation itself.

It should be noted that clause 20(2)(a) applies only to 'any part of the Permanent Works'. It would seem that if the employer uses the contractor's temporary works as well he might – e.g. a temporary road bridge – the responsibility for care of the works remains fully with the contractor.

Damage during defects correction period

Under clause 20(1)(c) the contractor takes full responsibility for outstanding works he has undertaken to finish, by clause 48(1), in the defects correction period. Under clause 20(3)(a), the contractor is responsible for any damage he causes to the rest of the works whilst undertaking the outstanding works or searching for defects.

Design faults

Under clause 20(2)(b) any damage due to a fault, defect, error or omission in the design of the works is within the category of excepted risk. However, design by the contractor is excluded.

The wording of the clause appears narrower in scope than that for an 'exception' in clause 22 for damage to persons and property.

There the phrase 'damage which is the unavoidable result of the construction of the Works' is used.

7.5 Clause 21 – insurance of the works

Clause 21 requires the contractor to insure the works against loss or damage in the joint names of the contractor and the employer. The purpose of this is to ensure that the contractor has funds to complete the works in the event of damage and alternatively to protect the employer's investment in the works.

The effect of putting insurance in joint names is to permit the employer to claim directly on the policy but it also has the advantage of preventing the insurer from exercising rights of subrogation between the parties.

Replacement at cost plus 10%

By clause 21(1) the insurance is to cover 'full replacement cost' plus an additional 10% for incidentals including fees, demolition and the like. The Fifth edition required insurance of the works 'to their full value'.

The change may be far more significant than the 10% addition. Replacement cost suggests an amount fixed by reference to actual cost incurred whereas value is more likely to be fixed by reference to bill rates.

Contractor's equipment omitted

The Sixth edition omits the requirement in the Fifth that the contractor should insure the construction plant to its full value; but note that insurance of the 'Works' covers both the permanent works and the temporary works.

Extent of cover

Clause 21(2) regulates the extent and duration of cover. It must:

(a) cover all loss or damage from whatsoever cause arising other than the excepted risks
(b) run from the works commencement date to the date of issue of the relevant certificate of substantial completion
(c) cover loss or damage arising during the defects correction period from a cause arising prior to the issue of any certificate of substantial completion
(d) cover loss or damage caused by the contractor in carrying out any work during the defects correction period.

Defective workmanship

Clause 21(2)(c) usefully confirms a point which is sometimes overlooked – that the contractor is not normally required to insure against repair or reconstruction of workmanship not in accordance with the specification. The clause does continue to say 'unless the Bill of Quantities shall provide a special item for this insurance' but this perhaps ignores the practical difficulties of obtaining such insurance.

As a general rule insurance against bad workmanship is not available. Thus if the contractor casts a concrete bridge abutment and it is rejected because of poor finish, the contractor must repair at his own expense. Of course, if the contractor's dump truck hits the abutment and causes damage that can be an insurance claim. The test for an insurance claim is usually damage not defect.

Unrecovered losses

Clause 21(2)(d) is a new provision which emphasises the point in clause 21(1) that insurance does not limit the obligations of the parties. It states that amounts not insured or recovered from insurers, whether by excesses or otherwise, shall be borne by the parties in accordance with their respective responsibilities.

This clause needs to be read in conjunction with clause 25 (evidence and terms of insurance) which, amongst other things, allows for controls on excesses.

7.6 Clause 22 – damage to persons and property

Clause 22 which relates to third party claims has no changes of substance from the Fifth edition but its layout is improved.

Clause 22(1) requires the contractor to indemnify the employer against losses and claims in respect of:

(a) death or injury to any person, and
(b) loss or damage to any property other than the works

arising out of or in consequence of the execution of the works.

The indemnity is to cover all claims, proceedings etc. and all charges and expenses whatsoever. It extends, therefore to legal costs as well as to sums claimed.

However on this point the case of *Richardson* v. *Buckinghamshire County Council* (1971) under the ICE Fourth edition is worth noting. A motor cyclist sued the County Council for injuries alleged to have been sustained when he fell off his machine at roadworks. The County Council successfully defended the claim but had to meet their own costs because the motor cyclist was legally aided. They tried to recover their costs from the contractor under clause 22 but it was held by the Court of Appeal they had not arisen 'out of or in consequence of the execution of the works'.

Exceptions

As with clause 20 there are exceptions to the general obligation. These are the items listed in clause 22(2) as being the responsibility of the employer:

(a) crop damage
(b) use or occupation of land
(c) right of the employer to construct on land
(d) damage which is the unavoidable result of the construction of the works
(e) claims arising from negligence or breach by the employer or his agents.

A small point to note is that wording of the final exception as it now appears in clause 22(2)(e) omits the reference to the engineer included in the corresponding clause of the Fifth edition. This may be because the engineer is seen as the employer's agent, but if the change goes deeper than this there are implications for both parties and the engineer.

Unavoidable damage

The exception in clause 22(2)(d) for damage which is the unavoidable result of the construction of the works is by far the most commonly invoked and the most contentious exception. Usually it is claimed for damage to property through excavations, dewatering or vibrations.

It is very much in the contractor's interest to make such a claim because:

(a) if he accepts responsibility and passes the matter on to his insurers for settlement his insurance premiums will reflect the claim the following year
(b) if there has been delay in the execution of the works he is liable for liquidated damages unless he can prove the cause to be an 'exception'

(c) the costs of rectifying damage may be recoverable under instructions issued by the engineer.

The burden of proving that damage was the unavoidable result of construction of the works rests with the contractor on the principle that he who asserts must prove.

The possibility of 'unavoidable' damage resulting from contractor's design raises some interesting questions. It cannot be intended that the employer should indemnify the contractor in such a situation but it may be that is what the contract provides. See the comment below on clause 22(4)(b).

Note that the exception in clause 22(2)(d) for unavoidable damage relates only to damage. It does not cover injury to persons.

Indemnity by employer

By clause 22(3) the employer is required to indemnify the contractor against all claims etc. made in respect of the exceptions.

Shared responsibility

Clause 22(4) on shared responsibility contains two separate provisions: one for the contractor; the other for the employer.

In clause 22(4)(a) the contractor's liability to indemnify the employer is reduced in proportion to the extent that the employer or his agents may have contributed to the damage or injury.

In clause 22(4)(b) the employer's liability to indemnify the contractor in respect of the exceptions in clause 22(2)(e) is reduced in proportion to the extent that the contractor or his agents etc. may have contributed to the damage or injury.

There is no reduction of employer's liability in respect of the other excepted risks in sub-clauses 22(a), (b), (c) and (d) – presumably on the basis that there can be no element of contractor's fault.

7.7 Clause 23 – third party insurance

Clause 23 provides for insurance against third party risks in much the same way that clause 21 provides for insurance of the works.

There are a number of changes from the Fifth edition:

(a) The phrase 'throughout the execution of the Works' which introduced clause 23(1) in the Fifth is omitted. No duration time is stated in the Sixth.
(b) Clause 23(1) requires the contractor to insure in the joint names of the contractor and employer.
(c) Also in clause 23(1) the insurance is to cover the exception in clause 22(2) (e) – damage from acts of neglect of the employer.
(d) Clause 23(2) requires the policy to have a cross liability clause such that the insurance applies to the contractor and employer as separate insured.

Amount of insurance

As in the Fifth edition the insurance shall be for at least the amount stated in the appendix to the form of tender.

7.8 Clause 24 – injury to work people

Clause 24 provides that the employer shall not be liable for damages or compensation for accidents or injuries to the contractor's workforce except to the extent that default of the employer is a contributory factor. The only change from the Fifth edition is that 'workman' has been replaced with 'operative'.

There is no express requirement for insurance cover but employers (in the broad sense) are required by statute to carry insurance against injury to their employees in the course of their employment.

There can be problems with labour-only and other self-employed persons who do not benefit from any employer's insurance cover. To guard against this and other claims clause 24 requires the contractor to indemnify the employer against all claims.

7.9 Clause 25 – evidence and terms of insurance

In the Fifth edition the control of insurances was split between clause 23(2) and clause 25. In the Sixth edition the various provisions have been consolidated into clause 25 with a number of minor changes.

Evidence and terms

Clause 25(1) requires the contractor to provide satisfactory evidence prior to the works commencement date that the insurances are in force. If required the policies must be produced for inspection.

The clause further provides that the terms of the insurance shall be to the approval of the employer and that the contractor shall, on request, produce receipts of payment of premiums. The requirement in the Fifth edition for insurance to be with an insurer approved by the employer has been dropped.

In many cases the employer will rely on the engineer to advise on insurance and to inspect and ensure that insurances are in force. This is a serious duty and engineers should note the case of *William Tompkinson & Son Ltd* v. *Church of St Michael* (1990) in which an architect was held to be negligent for failing to properly advise his client on the insurance provisions for a JCT Minor Works contract.

Excesses

Clause 25(2) provides that any excesses on policies shall be as stated by the contractor in the appendix to the form of tender. This is a new provision.

The provision is intended to guard against the contractor taking on insurance with high excesses and thereby leaving the employer exposed to the amount the excess in the event of the contractor's financial failure. Some employers state in the tender documents the maximum excesses they are prepared to accept.

Failure to insure

Clause 25(3) allows the employer to take out insurance when the contractor fails to produce satisfactory evidence that his insurances are in force. The premiums paid by the employer can be deducted from monies due to the contractor or recovered as a debt.

Contractors would be most unwise to allow this state of affairs to develop since the employer would insure in his own name leaving the contractor exposed to claims from the insurer under his rights of subrogation.

Compliance with policy conditions

Clause 25(4) contains another new provision requiring both the employer and the contractor to comply with the conditions laid down in the insurance policies and indemnify each other against any claims arising from failure to comply.

The effect of this is that if one party renders the insurance policies void by his actions or omissions, he then stands in the place of the insurer in providing cover.

Chapter 8

Statutes, streetworks, facilities and fossils

8.1 Introduction

This chapter deals with clauses 26 to 35 inclusive: a miscellaneous set of provisions concerned principally with statutes, streetworks, facilities and fossils.

The changes of substance from the Fifth edition are as follows:

(a) Clause 34 is no longer used. It previously required the contractor to pay wages and observe conditions not less favourable than those in the Working Rule Agreement. The Sixth edition has no such provisions.
(b) Clause 28 on patent rights has an additional indemnity from the employer to the contractor in respect of specified products.
(c) Clause 29 on noise and pollution now requires the employer to indemnify the contractor where the nuisance is the unavoidable consequence of carrying out the works.

Various other changes in detail have been made but disappointingly many obscure points of drafting remain from the Fifth edition.

8.2 Clause 26 – notices, fees and statutes

Clause 26 remains much the same as the Fifth edition but note the substitution of 'Contract' for 'Drawings Specifications' in clause 26(3) (b) and the deletion of 'specified' in clause 26(3) (c) in relation to temporary works.

Notices and fees

Clause 26(1) provides that the contractor shall give all notices and pay all fees required by statute, bye-laws or otherwise relating to the construction and completion of the works. This is subject to the provision in clause 27 for the employer to serve certain streetworks notices; the provision in clause 26(3) that the contractor is not generally responsible for obtaining planning permissions; and the provision in clause 26(2) for the employer to repay certain fees.

Notices may be required in respect of Building Regulations, health and safety matters, control of waste and waste disposal, the installation of services, demolitions and the use of explosives, use and temporary closure of highways, skips, adjoining properties and various other aspects of construction.

Repayment of fees

Clause 26(2) provides that the employer shall allow or repay the contractor:

(a) such sums as the engineer shall certify to have been properly payable 'and paid', and,
(b) rates and taxes paid by the contractor in respect of the site or any temporary structures situated elsewhere.

This is one clause which could have done with some clarification in its drafting but the wording remains identical to that in the Fifth edition. One question is – does the phrase 'any temporary structures situated elsewhere' oblige the employer to pay the rates on the contractor's site compound? Another is – does the phrase 'any structures used temporarily and exclusively for the purposes of the Works' oblige the employer to pay the rates on permanent buildings used by the contractor if they are given over to the contract? A third is – could either phrase apply to a special casting yard or similar set up outside the site by the contractor or the temporary exclusive use of a quarry?

The test for repayment in the clause is broadly expressed as 'used exclusively for the purposes of the Works'. This may or may not include administrative as well as constructional activities but it does not obviously exclude facilities used by the contractor of his own commercial choosing. If it is intended that the employer should only pay the rates on land and premises which the contractor cannot avoid using then the test could have been better expressed. As it is, contractors have nothing to lose by claiming repayment of rates and in fact many engineers and employers accept their entitlement.

Conforming with statutes

Clause 26(3) serves the general policy of the Conditions, which is to avoid reference to specific statutes by making a general provision that the contractor shall ascertain and conform with all statutes, regulations etc.

Employer to be indemnified

Clause 26(3) also requires the contractor to indemnify the employer against all penalties and liabilities of every kind for failure to conform with statutes except for the consequences of breach which is the unavoidable result of complying with the contract or the engineer's instructions.

There may well be a legal difference between the exception here 'the unavoidable result of complying with the Contract' and the exception in clause 22(2)(d) 'the unavoidable result of the construction of the Works'. The first, it is thought, is a stricter test which would not include the consequences of the contractor's chosen method of working.

Contractor not relieved of legal responsibility

The exception to the indemnity given to the employer does not, of course, relieve the contractor of his legal responsibility for any breach of statute or regulation. All that it does is relieve the contractor from his indemnity to the employer.

Notwithstanding the firmly stated requirement in clause 13(1) that the contractor shall comply with and strictly adhere to the engineer's instructions, the contractor has obligations at statute and common law which must take priorty.

The case of *Eames* v. *North Hertfordshire District Council* (1980) is instructive. In that case a contractor erected a portal frame on made-up ground. Both the architect and the Council's building inspector allowed the work to proceed. Nonetheless the contractor was held liable for breach of a statutory duty (under the Building Regulations) and liable in negligence. Judge Fay QC referred in his judgment to the comment of Lord Wilberforce in *Anns* v. *London Borough of Merton* (1978):

> 'Since it is the duty of the builder, owner or not, to comply with the bye-laws, I am of the opinion that an action could be brought against him in effect for breach of statutory duty by any person for whose benefit or protection the bye-law was made.'

Engineer to issue instructions

By clause 26(3)(b) the engineer is required to issue instructions, including any variation under clause 51, to ensure conformity of the contract with statutes and regulations etc.

It is suggested that if the contractor has any doubt on the legality of what he is doing he should seek professional advice and refer the matter promptly to the engineer.

Planning permission

Clause 26(3)(c) provides the major exception to the contractor's obligation to give all notices and pay all fees in stating that the contractor is not responsible for obtaining planning permission.

The Sixth edition refers to permission 'necessary in respect of the Permanent Works or of any Temporary Works design supplied by the Engineer'. The Fifth edition said 'necessary in respect of the Permanent Works or any Temporary Works specified or designed by the Engineer'. The change suggests that the contractor may now be responsible for obtaining planning permission for temporary works 'specified' but not designed by the engineer. This might, for example, apply to a requirement for a prominent site notice board.

The statement in clause 26(3)(c) that the employer warrants that all permissions have been or will be obtained in due time gives the contractor good grounds for claim if there is any delay. The Canadian case of *Ellis-Don Ltd* v. *Parking Authority of Toronto* (1978), much quoted as authority on winter working and the recovery of overheads, concerned the failure of the Authority to obtain building permits in time.

8.3 *Clause 27 – public utilities streetworks*

The drafting committee of the Sixth edition have announced that, when the new Streetworks Bill becomes law, it is their intention to issue a revision to clause 27. They propose to examine the possibility of greatly shortening the clause and eliminating direct reference to statute.

The Public Utilities Streetworks Act 1950, which presently applies, sets out the framework for highway authorities, statutory undertakers and the like to be given notice of proposed works which may affect their interests and it provides for protection of those interests.

Clause 27 as it currently stands is identical to its predecessors in the Fifth and Fourth editions. It can be summarised as follows:

(a) Clause 27(1) defines the Act and other expressions.
(b) Clause 27(2) requires the employer to notify the contractor before the commencement of the works of the legal status of land within the works.
(c) Clause 27(3) requires the employer to serve all notices due under the Act.
(d) Clause 27(4) requires the contractor to give the employer 21 days' notice before commencing relevant works.
(e) Clause 27(5) requires the contractor to give a repeat notice if he fails to commence within two months.
(f) Clause 27(6) deals with delays and additional costs resulting from variations which involve the Act.
(g) Clause 27(7) requires the contractor to comply with his obligations under the Act and indemnify the employer against his default.

8.4 *Clause 28 – patent rights and royalties*

Clause 28(1) covering breach of patent or copyright has been significantly amended so that the indemnity from the contractor to the employer, which in the Fifth edition covered everything, now in the Sixth edition excludes infringements resulting from compliance with the design or specification.

In respect of such infringements the employer must now indemnify the contractor.

Clause 28(2) simply requires the contractor to pay royalties, rent and other payments for getting stone, sand, clay etc. for the works.

This is a statement of the obvious and its inclusion in the Conditions can only be explained in relation to some provision by the employer of sources of materials.

On the face of it, rates for any quarry, barrow pit etc. would be the contractor's responsibility but consider the opening words of the clause 'Except where otherwise stated' and the comments earlier in this chapter on rates under clause 26.

8.5 *Clause 29 – interference, noise and pollution*

Clause 29 deals with various aspects of nuisance to third parties as opposed to damage, which is dealt with in clause 22. The first part of the clause deals with traffic, the second part with noise and other pollution.

The only change of significance from the Fifth edition is that the employer now indemnifies the contractor for noise, disturbance or other pollution which is unavoidable.

Traffic

Clause 29(1) requires that all operations for the construction of the works shall be carried out so as not to inferfere 'unnecessarily or improperly' with:

(a) the convenience of the public
(b) access to roads, footpaths, properties
(c) use of roads, footpaths, properties.

Indemnity

The contractor is required to indemnify the employer against all claims and proceedings.

There appears to be a discrepancy between the obligation on the contractor to avoid interference and the indemnity he must give. The obligation is qualified by the phrases:

> 'so far as compliance with the requirements of the Contract permit' and
> 'not to interfere unnecessarily or improperly'.

The indemnity is not so qualified.

Counter-indemnity

It also seems odd that the counter-indemnity by the employer given in the new clause 29(4) applies only to noise, disturbance or pollution and does not include interference.

Noise, disturbance and pollution

Clause 29(2) requires that all work shall be carried out without unreasonable noise or disturbance or other pollution.

Clause 29(3) requires the contractor to indemnify the employer against claims and proceedings except to the extent that noise, disturbance or pollution are the 'unavoidable consequence of carrying out the Works'. In the Fifth edition the contractor's indemnity was absolute.

Clause 29(4) is the new requirement for the employer to indemnify the contractor against claims which are so unavoidable. It is expected that this will prove to be a change of some importance to engineers, employers and contractors.

Engineers will have to recognise at the design stage and when supervising construction that the employer is at risk for nuisance which is unavoidable. There is, of course, great scope for dispute here since contractors who have previously automatically passed on claims to their insurers will now surely, just as automatically, pass on claims to the employer alleging unavoidability.

Off-site liability

When the contractor was fully responsible for nuisance it mattered not whether it arose on or off the site but the question must now be asked – is the employer potentially liable for nuisance arising off the site? What for example is the position on nuisance from a borrow pit? It may be that the employer is now responsible for unavoidable nuisance whether or not the engineer acts under clause 1(1)(v) to designate the borrow pit as part of the site. The engineer, however, would be well advised to take a cautious approach to any extension of the site.

8.6 Clause 30 – damage to highways

Clause 30 is intended to settle, as between the contractor and the employer, their respective liabilities under the Highways Act 1980. The Act allows the highway authority to recover its excess maintenance expenses arising from excessive weight or extraordinary traffic. Broadly the scheme is that the contractor takes all risk from movement of his equipment and temporary works and the employer takes the risk from materials and manufactured articles. There is no change of any substance from the Fifth edition.

Clause 30(1) requires the contractor to use every reasonable means to prevent extraordinary traffic and to use suitable routes and vehicles to ensure that any inevitable extraordinary traffic is limited as far as possible and no unnecessary damage is caused to roads and bridges.

Clause 30(2) provides that the contractor must pay the costs of strengthening or improving the roads or bridges to facilitate movement of his equipment or temporary works and must indemnify the employer against all claims.

Clause 30(3) requires the contractor to notify the employer as soon as he becomes aware of any damage caused by the transport of materials or manufactured articles. The employer is to negotiate and settle any claims and to indemnify the contractor. If the claim is due to any failure by the contractor to use reasonable means and limit damage as required by sub-clause (1) then the engineer is required to certify the amount due from the contractor to the employer.

As in clause 20, this brings the engineer into the field of the loss adjuster. Almost invariably insurance companies are involved in claims for damage and they are not disposed to take a back seat while employers negotiate and engineers apportion their liabilities.

8.7 Clause 31 – facilities for other contractors

Clause 31 makes clear that the contractor is not entitled to sole possession of the site. It remains broadly unchanged from the Fifth edition except for the addition of profit in claims for additional work in conformity with the same change elsewhere in the Conditions.

Clause 31(1) obliges the contractor to afford all reasonable facilities, in accordance with the requirements of the engineer or the engineer's representative to:

(a) other contractors employed by the employer and their workmen

(b) workmen of the employer
(c) workmen of authorities or statutory bodies employed in the execution of work not in the contract
(d) workmen of authorities or statutory bodies employed on any contract entered into by the employer in connection with or ancillary to the works.

Engineer's representative

The inclusion of the engineer's representative in defining requirements is new. It allows him to act without delegated powers on this issue. Although his functions are to watch and supervise, note the wording of clause 2(3)(b): 'Nor as expressly provided hereunder to order any work involving delay or any extra payment.' Clause 31(1) is one of the few clauses giving the engineer's representative such express power.

Statutory undertakers

The references in clause 31(1) to authorities and statutory bodies are probably wide enough to avoid disputes of the kind in *Henry Boot Ltd* v. *Central Lancashire Development Corporation* (1980).

In that case there was an extension of time dispute on a building contract as to whether statutory undertakers, who were laying mains to a building site, were engaged in executing work not forming part of the contract or were carrying out the work in pursuance of their statutory obligations. The decision was that they were involved in the former because they were carrying out the work not because statute obliged them to do so but because they had contracted with the employer to do so.

Reasonable facilities

The phrase 'reasonable facilities' in clause 31 is open to a variety of interpretations. It probably means no more than allowing access and space in which to work. It may also include the use of scaffolding, messrooms, toilets and the like and possibly the supply of power and water.

The uncertainty of the phrase begs the question: what if the contractor complains that the requirements of the engineer are to afford 'unreasonable facilities'? The answer would seem to be that the contractor must comply with any instructions and seek reimbursement through clauses 31(2) or 13(3) as appropriate or even clause 51 if a variation is involved.

Delay and extra cost

Clause 31(2) entitles the contractor to an extension of time and recovery of cost for any delay and cost beyond that reasonably to be foreseen at the time of tender. This is the same test for recovery as in clause 13.

Claims in respect of statutory undertakers

Not surprisingly since clause 31 specifically refers to statutory undertakers, clause 31(2) is often taken as the starting point for claims for delays caused by statutory undertakers. More often than not the dispute is on what could reasonably have been foreseen at tender. That will depend on what information the contractor was given at the time of tender on the estimated duration of the statutory undertaker's works and the facilities to be provided.

If the engineer gave duration times which were inadequate and nothing was specified on facilities, the contractor starts with a good case.

However, clause 31(2) is not the only, or even the best, provision for the contractor's recovery for statutory undertakers' delays. Under clause 42(2) for example the employer is required to give possession of the site as required by the accepted clause 14 programme and many delays caused by statutory undertakers can be seen as depriving the contractor of his rights of possession under that clause. Any failure of possession which involves the contractor in delay or extra cost entitles the contractor to an extension of time and recovery of cost without consideration of 'that reasonably to be foreseen' as in clause 31.

8.8 Clause 32 – fossils

As befits its title, clause 32 on fossils has remained unchanged through the Fourth, Fifth and Sixth editions.

It deals with ownership of things of interest found on the site only as between the employer and the contractor and they are deemed to be the absolute property of the employer. The contractor is required to take precautions to prevent removal or damage and is to acquaint the engineer immediately upon discovery. The contractor is to carry out at the expense of the employer the engineer's orders for disposal.

The absence of any reference to extension of time for delay and the uncertainty of what is meant by 'expense' are but modest deficiences in the drafting. The contractor should have little difficulty obtaining an extension under clause 12 (artificial obstructions), clause 13 (instructions), clause 42 (lack of possession), or the general 'special circumstances' of clause 44. Costs could follow either clauses 12, 13, 42 or 51 or could be on a daywork basis under clause 52(3).

8.9 Clause 33 – clearance of site

The obligation in clause 33 for the contractor to clear the site on completion can be inferred from both clause 8 (general obligations) and clause 53(3) – (disposal of contractor's equipment).

Completion is not a term defined in the contract but it is reasonable to expect the contractor to commence cleaning up on the issue of the certificate of substantial completion rather than delay to the end of the defects correction period.

8.10 Clause 35 – returns of labour and equipment

Clause 35 has been modified slightly so that either the engineer or the engineer's representative can now require returns of labour and the contractor's equipment. As in the Fifth edition, the provision extends to sub-contractors.

These are useful provisions for the engineer to use. The information obtained on resources can be relevant to claims for delay and extra cost and details of equipment are needed for clause 53(1) – (vesting of equipment).

Chapter 9

Workmanship, materials and suspensions

9.1 Introduction

This chapter covers the following clauses:

clause 36 – workmanship and materials
clause 37 – access to site
clause 38 – examination of work
clause 39 – removal of unsatisfactory work
clause 40 – suspension of work

Clauses 36, 37 and 38 are unchanged from the Fifth edition. Clauses 39 and 40 have minor drafting changes and clause 40 also has the new common reference to profit on additional work.

9.2 Clause 36 – workmanship, materials and quality

Clause 36(1) requires that 'all materials and workmanship shall be of the respective kinds described in the Contract and in accordance with the Engineer's instructions'.

The general rules on materials have been covered in section 2.6. Broadly it can be said that where the contractor is free to choose materials they must be of merchantable quality and fit for purpose; where the contractor is required to use particular products they must be of merchantable quality but the contractor does not warrant they are fit for purpose.

On workmanship, the general rule is that the contractor is to use reasonable skill and care. Thus section 13 of the Supply of Goods and Services Act 1982 says:

'In a contract for the supply of a service where the supplier is acting in the course of a business, there is an implied term that the supplier will carry out the service with reasonable skill and care.'

Where the contractor is required to both design and build, the general rule is that fitness for purpose applies.

The above general rules can, of course, be subject to modification by the terms of specific contracts; and in particular by the specified quality or specified performance required by the contract. Thus the contractor does not fulfil his obligations to supply concrete of a specified minimum crushing strength by supplying concrete of a lower grade and claiming it is of merchantable quality and fit for its purpose. Nor does the contractor fulfil his obligation to finish the surface of concrete to a specified standard by finishing to a lower standard and claiming he has used reasonable skill and care.

As to design and build, see the comments in section 6.2 on contractual provisions to relieve the contractor of his obligations for fitness for purpose.

'In accordance with engineer's instructions'

It is not fully clear what is meant by 'and in accordance with the Engineer's instructions' in the opening sentence of clause 36(1).

It is probably intended simply to cover work which is not detailed in the specification, bills of quantities or drawings so that it is supplemental to 'the Contract'. It is not thought to be a further power of the engineer to vary descriptions in the contract; he already has such power under clause 51 and the power to give instructions under clause 13.

Testing

Clause 36(1) provides for such tests as the engineer may direct at:

(a) the place of manufacture
(b) the site
(c) such other places as specified in the contract.

'Such other places' could include named testing laboratories or the engineer's or employer's premises.

The contractor is to supply assistance and equipment as is normally required for examination and testing; and is to supply samples 'before incorporation in the Works' as selected and required by the engineer.

Cost of samples

Clause 36(2) provides that all samples shall be supplied by the contractor at his own cost if:

(a) the supply is clearly intended by the contract, or
(b) the supply is provided for in the contract.

Otherwise, the cost of samples is to be met by the employer.

Note the distinction between 'Contract' as used in clause 36(2) and 'Contract ... and Engineer's instructions' in clause 36(1). Thus it is fairly clear that if the engineer orders samples beyond those required by the specification, bills of quantities or drawings, they are not in a narrow sense 'in the Contract' and must be paid for by the employer.

This is not always recognised by engineers who, not infrequently, assume the contractor is under a general duty to supply samples of all materials to prove that they meet the specification.

Cost of testing

Clause 36(3) sets out how the costs of testing are to fall between the contractor and the employer. The clause provides that the contractor shall bear the costs of tests which are:

(a) clearly intended by the contract, or
(b) provided for in the contract, or
(c) for load tests, particularised in the specification or bill of quantities in sufficient detail to enable the contractor to have allowed for the cost in his tender.

For any test not so intended, provided for or particularised the contractor is to bear the cost only if the tests show workmanship or materials not in accordance with the contract; if the tests show compliance with the contract, the employer bears the cost.

In short, the contractor pays for all tests which are 'specified' in the contract, whether they pass or fail; but for tests which are 'additional', the contractor pays for failures and the employer pays for passes. The principle of this may seem clear but in practice the outcome is often uncertain and potentially unfair.

Testing following failures

The problem with the basic rule in clause 36(3) is that failure in one element of the works may oblige the engineer on grounds of prudence, having regard to safety or contractual default, to order additional tests. Why, it may be asked, should the employer pay for such tests?

The short answer to this is that it seems to be the policy of the contract because similarity can be found in clauses 38 and 50. The alternative view is that, if testing under clause 36 can be regarded as a serial process, the costs of follow-on tests do not necessarily fall on the employer. Attractive as this proposition sounds it is some way from the actual wording.

It is worth noting that the draftsmen of the JCT 80 standard form of building contract have grasped this particular nettle. The employer is no longer liable for the costs of opening-up and testing where the instructions given are reasonable in the light of a code of practice issued for dealing with non-compliance.

Load tests

Finally, on clause 36(3) and the costs of tests provided for in the contract, it should not be taken from the wording of the clause that the contractor is only to allow in his tender for the costs of load tests. Clearly the contractor must allow for all tests which are specified; the reference to load tests seems to have been made only to avoid vague descriptions of expensive tests.

9.3 Clause 37 – access to site

Clause 37 expressly provides, what would in any event be implied, that the engineer and any person authorised by him shall at all times have access to the works and the site. The clause covers workshops and places of manufacture, and adds (presumably in relation to premises not owned by the contractor or the employer) that the contractor shall afford every facility and every assistance in obtaining access.

The practical effect of the clause is that it obliges the contractor to make available to the engineer some means of access to the various parts of the works.

9.4 *Clause 38 – examination of work*

Clause 38 deals with the examination of work before covering up and the uncovering of work for inspection.

Examination before covering up

Clause 38(1) contains the following provisions:

(a) no work to be covered up without the consent of the engineer
(b) the contractor to afford full opportunity to the engineer to examine and measure
(c) the contractor to give due notice when work is ready for examination
(d) the engineer to attend without unreasonable delay, unless he considers it unnecessary and informs the contractor accordingly.

These apparently simple and practical rules are not without difficulty in their application. Firstly, there is the question of the importance or the volume of the work itself and then there is the question of timing. Both matters are related to the level of supervision available to the engineer and both have some bearing on what is unreasonable delay.

The contractor may claim that anything which impedes continuity of progress is unreasonable delay; the engineer may reply that he can only be in one place at a time and some measure of delay is inevitable and is envisaged by clause 38(1).

In practice commonsense usually prevails and the contractor and engineer establish a working relationship often based on a formal notice procedure with written notices from the contractor and return slips from the engineer. Clause 38 does not actually require written notices but on a large contract they are indispensable.

In the event of dispute, two questions arise:

(a) Can the contractor proceed without waiting for the engineer?
(b) Is the contractor entitled to an extension of time and recovery of costs for unreasonable delay?

Covering up without consent

If the contractor does put work out of view without the consent of the engineer that on the face of it is a breach of contract for which clause 38(2) provides a remedy in the power of the engineer to order uncovering. However, if the contractor has given notice to which the engineer has not responded, and this is accepted as a regular occurrence or pattern of behaviour on the particular contract, then it may be reasonable to imply consent to proceed from the engineer's conduct, and there is no breach by the contractor.

Recovery of cost for delay in attending

It is important to note that strictly on the wording of clause 38(1) the contractor is prohibited from covering up without consent. It follows that he is obliged to stand and wait for the engineer to attend if notice has been given. It also follows that the contractor is entitled to his costs for unreasonable delay by the engineer in attending.

However, because there is no express provision for recovery of cost in the clause itself, the contractor must either base his case on breach of contract under clause 2(1)(a) or on a common law breach.

Extension of time for delay in attending

As for extension of time this, if granted for unreasonable delay, would seemingly have to come under 'other special circumstances' in clause 44. Such circumstances might apply if, for example, the delay was an infrequent occurrence or caused by, say, the unforeseen absence of a member of the engineer's supervising team. However, if the delay was part of a regular pattern it could not truly be said to be a special circumstance. The contractor might then fare better by claiming breach of contract in failure by the engineer to carry out his duties specified in the contract, clause

2(1) (a), and claiming, not for an extension of time, but time to be at large. See chapter 11 for further comment on this.

Uncovering and making openings

Clause 38(2) gives the engineer power to order the contractor to uncover, or make openings in, any part of the works.

The clause provides that if the parts have been covered after compliance with sub-clause (1) and found to be in accordance with the contract the cost shall be borne by the employer, but if otherwise by the contractor.

To be certain of payment the contractor needs to show that he has been given consent to cover up and the work is satisfactory. Neither on its own is sufficient. However, it is suggested that the contractor has a good case for payment if he can show that he gave notice, allowed a reasonable time before covering up and the work was satisfactory.

Extension of time

Providing the contractor can prove his entitlement to payment he should have no difficulty obtaining an extension of time for any delay caused by the uncovering and reinstatement. This will follow the engineer's instruction to uncover which can be attributed to clause 13.

'In accordance with the Contract'

In the comment on clause 36 it was noted that the phrase 'in the Contract and in accordance with the Engineer's instructions' was used. In clause 38(2), the phrase used is 'executed in accordance with the Contract'. An interesting debating point could arise if the engineer by instruction varied the specification so that the work uncovered, although in accordance with the engineer's instructions, was not in accordance with the 'Contract'. But that, perhaps, takes one back to the interpretation of the phrase in clause 36.

Consequential uncovering

The problem of apparent unfairness to the employer in being liable to pay for uncovering made in consequence of previously discovered defects applies to clause 38 as to clause 36.

However, the employer would have a better case for recovering his costs from the engineer on a negligence basis if clause 38, rather than clause 36, applied since either the engineer's consent or his inspection would seem to be deficient.

9.5 Clause 39 – removal of unsatisfactory work and materials

Clause 39 gives the engineer power to order the removal of unsatisfactory work and materials, the employer the right to act if the contractor is in default, and confirms that failure by the engineer to disapprove does not prevent later disapproval.

These express powers are valuable since it is not certain that either the engineer or the employer could find an implied term to justify intruding into the contractor's performance prior to completion.

Many commentators on the ICE Conditions have suggested that to be of full practical effect the provisions should go beyond their negative approach and should empower the engineer, with the employer's consent, to accept sub-standard work – albeit at a reduced price. For further comment on this see chapter 13 on variations and in particular the case there of *Simplex Concrete Piles* v. *St Pancras Borough Council* (1958).

Removal of work and materials

Under clause 39(1) the engineer can, during the progress of the works, instruct the contractor in writing to:

(a) remove from site within specified times materials which in the opinion of the engineer are not in accordance with the contract
(b) substitute with materials in accordance with the contract
(c) remove and re-execute, notwithstanding previous testing or payment, work which in the opinion of the engineer is not in accordance with the contract either by:

  (i) materials and workmanship, or
  (ii) design by the contractor.

'During the progress of the works'

This phrase suggests that the engineer's powers under clause 39(1) are not exercisable during the defects maintenance period.

'In the opinion of the engineer'

The question is what does the contractor do if he disagrees with the opinion of the engineer? Does he follow the instruction as he is required to do by clause 13 and seek to recover his costs by a claim or arbitration or has he other alternatives?

Firstly the contractor should consider if it is the engineer himself or the engineer's representative with delegated powers who has given the instruction. If it is the latter the contractor can refer the matter to the engineer under clause 2(7). For the logic of this, note the reference in clause 39(3) to 'any person'. If no change is forthcoming the contractor has little option but to comply first and claim later. Any alternative course of action risks determination under clause 63 or direct action by the employer under clause 39(2).

'Notwithstanding any previous test'

The obligation of the contractor is to construct the works in accordance with the contract – passing tests is incidental. Hard as it may seem on the contractor, if work is found to be not up to standard it is the standard and not the previous test which takes precedence.

'Notwithstanding interim payment'

Clause 60(8) gives the engineer express power to correct certificates and there is nothing in the payment process to suggest finality in approval.

'Design by the contractor'

Clause 39(1) (c) (ii) is new. It follows the introduction into the Sixth edition of limited use of contractor's design.

The effect is to widen the scope of clause 39 in respect of the contractor's design work so that the engineer can form an opinion not merely on workmanship and materials but also on design. Thus if the engineer finds part way through the works that the contractor's design is faulty he now has power to order the removal of the work built notwithstanding that its workmanship may be perfectly satisfactory.

Default in compliance

Under clause 39(2) the employer is entitled to employ and pay others to carry out the instructions of the engineer and to recover the costs from the contractor if the contractor defaults on complying with those instructions.

Note also, however, that under clause 63(1)(b)(iii) if the contractor fails to remove goods or materials or pull down and replace work for 14 days after receiving notice to do so then procedures to determine the contractor's employment can be commenced.

Both are powerful remedies and although the latter is the more extreme it is perhaps the more practical because if the employer was to enter the site, demolish work, and reconstruct with the contractor still in occupation and proceeding, it might almost lead to a breach of the peace.

The two provisions do not have any procedural link and appear to stand independently. Thus clause 39(1) allows the engineer to set times for compliance whereas clause 63(1)(b)(iii) sets a 14 day deadline for compliance.

In clause 62 there is an additional provision for the employer to act in respect of urgent repairs but this operates only in respect of accidents or failures and it is not applicable to merely unsatisfactory work.

Failure to disapprove

Clause 39(3) causes contractors as much dismay as any clause in the contract. It expressly provides what can in any event be gathered elsewhere, that failure of the engineer or any of his assistants to disapprove work does not prejudice his power to subsequently take action under the clause.

That is to say, work which has been once approved can later be rejected. This is not as unfair as it seems. The contractor's obligation is to construct the works in accordance with the contract and that is what the employer is entitled to receive. The engineer supervises on behalf of the employer, not the contractor, and he is not empowered to relieve the contractor of any of his obligations under the contract. If the engineer misses a fault or makes a mistake he is not only empowered to correct it, he has a duty to do so.

The contractor has no obvious redress against the effects of reversals of approvals which can be immense in both cost and delay. He can refer the matter to the engineer under clause 2(7) or under clause 66 for a further reversal but if that fails he must comply with instructions and then claim if he sees fit and has adequate grounds.

There have been instances where contractors have catalogued a series of late rejections and reversals of approvals on a particular contract and have claimed against the employer on the general ground that unreasonable behaviour by the engineer is a breach of contract. The difficulty is proving unreasonable behaviour but if it can be sustained the claim cannot lightly be dismissed.

'Any person acting under him'

The wording of clause 39(3) is most unusual in its reference to both the engineer and to 'any person acting under him'. This suggests that persons other than the engineer have powers under clause 39. But that is clearly not so. The clause empowers only the engineer to act.

The engineer may of course delegate his powers under clause 39 as he can with most of the other clauses of the contract. But throughout the contract persons acting with delegated powers still come under the reference of the engineer.

What then is to be made of the reference to this other 'person'? The suggestion offered is that powers of delegation operate slightly differently under this clause than elsewhere. Thus if the resident engineer has delegated powers and fails to reject such work then the engineer can, without reversing that delegation, himself reject work.

9.6 *Clause 40 – suspension of work*

Clause 40 gives the engineer power to suspend work in specified circumstances and the contractor rights of determination if the suspension lasts for more than three months. The only change from the Fifth edition is the addition of profit on additional work.

Legal position

It is generally thought that without an express provision for suspension neither party can suspend, nor can the engineer order suspension. To suspend without express provision amounts either to breach or reliance on an implied term. The difficulty with the implied term is that an assumption has to be made to justify the action but the courts will only agree if it is necessary to give business efficacy of the contract. The classic explanation of this was given by Lord Pearson in *Trollope & Colls Ltd* v. *North West Metropolitan Regional Hospital Board* (1973) when he said:

> 'The court will not even improve the contract which the parties have made for themselves, however desirable the improvement might be. The court's function is to interpret and apply the contract which the parties have made for themselves. If the express terms are perfectly clear and free from ambiguity, there is no choice to be made between different possible meanings; the clear terms must be applied even if the court thinks some other terms would have been more suitable.
>
> An unexpressed term can be implied if and only if the court finds that the parties must have intended that term to form part of their contract: it is not enough for the court to find that such a term would have been adopted by the parties as reasonable men if it had been suggested to them: it must have been a term that went without saying, a term *necessary* to give business efficacy to the contract, a term which, though tacit, formed part of the contract which the parties made for themselves.'

In the case of *Canterbury Pipe Lines Ltd* v. *Christchurch Drainage Board* (1979), in a dispute over amounts certified by the engineer on a drainage contract, the contractor suspended work which led to the employer taking the work out of the contractor's hands. It was held that the contractor had no right to suspend, and in commenting but not ruling on the general right of the contractor to suspend Mr Justice Cooke said this:

> 'We are against recognising such a right when the architect or

engineer has declined to issue the certificate stipulated for by the contract as a condition precedent to the employer's duty to pay. It would disrupt the scheme of these contracts. It could encourage contractors to take the law into their own hands. They might stop work which in the public interest needs to be done promptly. In such cases, if the contractor cannot or does not wish to rescind and cannot prove impossibility or its equivalent, he will be left with whatever remedies regarding the recovery of progress payments may be available to him under the contract. As already indicated we do not exclude the view that a case of impossibility or its equivalent *might* be made out by proving that for want of money the contractor could not carry on or could not reasonably be expected to do so.'

So much then for the position of the contractor taking action into his own hands. As for the position of the engineer or the employer suspending work without an express provision that would almost certainly amount to prevention and a breach of contract.

Simply expressed, the principle of prevention is that one party cannot impose a contractual obligation on the other party where he has impeded the other in the performance of that obligation.

Nature of clause 40

No clause in the contract needs to be more carefully read than clause 40. It does not oblige the engineer to do anything; the specified events are not grounds for suspension but are exceptions to rights of payment; if the suspension lasts for more than three months the contractor has rights of determination even if the employer is blameless. It gives no right whatsoever for the contractor to suspend work on his own volition.

What the clause does is give the engineer discretionary power to suspend work. He can do that on whatever grounds he considers necessary. It may be that he can act under the instructions of the employer as well as on his own professional judgment.

Suspension of work by the engineer is not a breach of contract; it is provided for in the contract. Clause 40 provides compensation to the contractor in that he is entitled to recover cost plus extension of time, unless the suspension results from his default, weather or safety; and in the case of prolonged suspension he has rights of determination.

Order and procedure

The procedural provisions of clause 40(1) are:

(a) the engineer gives a written order to suspend the progress of the works – or any part thereof – for such time or times – and in such manner – as the engineer considers necessary, and
(b) the contractor shall properly protect and secure the work, so far as is necessary in the opinion of the engineer.

'As the engineer may consider necessary'

If there are any grounds for doubting that the engineer has unfettered grounds to suspend or that he can do so acting under instructions from the employer – who may perhaps have temporarily run out of funds or be rethinking his need for the project – they may arise from the phrase 'the Engineer may consider necessary' in the first sentence of clause 40(1). However it is suggested that the phrase attaches principally to 'such manner' and is meant to have practical not legal effect.

Payment

Under clause 40(1) the contractor is to be paid, subject to clause 52(4) which relates to notice, the extra cost of giving effect to the engineer's order and instructions unless the suspension is:

(a) provided for in the contract
(b) necessary by reason of weather conditions
(c) necessary because of contractor's default
(d) necessary for proper execution of work
(e) necessary for safety.

Profit is to be added to cost in respect of any additional permanent or temporary work.

Other provisions

The exception in clause 40(1) (a) – 'otherwise provided for in the Contract' – would seem to relate to specific lack of possession, or to closure periods or the like detailed in the specification, bills, or drawings since there are no 'other' provisions in the Conditions for suspension.

Weather conditions

Suspension for weather conditions is usually given at the contractor's request, since although there is no recovery of cost, there is usually entitlement to extension of time. The position when the engineer suspends for weather conditions against the contractor's wishes is potentially contentious. The engineer will probably have in mind that temperatures are too low or too high for compliance with specifications or that working in wet weather or snow will render materials unsuitable. The contractor will often contend that he proceeds at his own risk and that the engineer has no right to interfere. However, the fact is that by clause 40(1) the engineer has a right to interfere and it is suggested he has a duty to the employer to prevent needless destruction of materials where the costs of replacement fall on the employer.

Contractor's default

The phrase 'by some default on the part of the Contractor' in clause 40(1) (b) seems to be wide enough to cover both practical and administrative defaults. The question is sometimes asked – can the engineer suspend the works if the contractor fails to provide a programme as required by clause 14? The answer is 'Yes', provided the engineer considers it necessary. All that is in dispute is whether or not the contractor should be paid. Where the contractor is clearly in default he may have some difficulty overcoming clause 40(1) (b).

Proper execution – clause 40(1) (c)

If the engineer suspends to achieve proper execution the contractor cannot expect to be paid. Common examples would be persistent failure to clean brick ties or to properly vibrate concrete.

Safety – clause 40(1) (c)

Notwithstanding the provisions of clauses 8(3) and 19(1) on the contractor's responsibility for safety the engineer has a statutory duty to act when he sees danger in work in which he is involved. Clause 40 gives him the power to act and he could be negligent in not using that power.

Shared responsibility

Clause 40(1) (c) recognises that a suspension for proper execution or safety could arise from some default of the engineer or employer or could be an excepted risk under clause 20(2), of which use by the employer or design fault are the most obvious. The exception to the contractor's right to payment is modified accordingly in such circumstances.

Extension of time

Under clause 40(1) the contractor is entitled to an extension of time for completion for delay caused by a suspension, unless the suspension is:

(a) otherwise provided for in the contract, or
(b) necessary by reason of some default of the contractor.

The general test for entitlement as set out in clause 44 is where 'the delay suffered fairly entitles' the contractor to an extension. Therefore a suspension to a non-critical activity will not justify an extension. The contractor can, however, get an extension for a suspension for weather conditions provided it meets the general test.

The phrase in brackets in the final paragraph of clause 40(1) 'including that arising from any act or default of the Engineer or the Employer' is interesting in that it expressly covers prevention rather than leaving it to be implied. For further comment on this see under 'special circumstances' in chapter 10.

Suspension longer than three months

Clause 40(2) deals with a suspension which has not been lifted within 3 months. Unless the suspension is either:

(a) provided for in the contract, or

(b) necessary by reason of contractor's default

the contractor may give written notice requiring permission to proceed within 28 days.

If the engineer does not grant such permission the contractor may, by further written notice:

(a) treat suspension of part of the works as a variation under clause 51, or
(b) treat suspension of the whole of the works as abandonment of the contract by the employer.

The practical and financial consequences of the contractor taking either action are potentially such that the engineer would be well advised to notify the employer at the start of the 28 day period of the circumstances so that instructions and legal advice can be obtained in good time. The clause leaves many questions unanswered.

Abandonment

Firstly, on the most serious matter of abandonment, which it should be noted is the only provision in the Conditions giving the contractor a right of determination, no procedures and rights equivalent to those in clause 63 are specified. It is suggested that the contractor would have all the rights of common law determination and would not be restrained by the concept of 'determination of his employment'. For further comment on this see chapter 19.

Omission variation

The question which needs to be asked of an assumed omission variation for a prolonged suspension of part of the works is: could the engineer reinstate the omitted part by later variation? There is no obvious contractual barrier to this but the contractor would, of course, be entitled to have both variations valued under clause 52, that is the omission and the reinstatement.

Finally there is the significance of the phrase in brackets in clause 40(2) 'but is not bound to'. Is this intended to emphasise that the contractor may exercise patience and await lifting of the suspension or is it to emphasise that the contractor can wait indefinitely after the 28 day period has elapsed before choosing to exercise his determination rights?

Chapter 10

Commencement, time and delays

10.1 Introduction

This chapter covers:

clause 41 – works commencement date
clause 42 – possession of site
clause 43 – time for completion
clause 44 – extension of time for completion
clause 45 – night and Sunday work
clause 46 – rate of progress.

Clause 45 remains identical to its counterpart in the Fifth edition; clause 43 has new terminology but no change of substance; the other four clauses have significant changes, mainly directed at improvements in management and administration.

10.2 Clause 41 – Commencement

Clause 41(1) states that the commencement date shall be the date specified in the appendix or, if none is specifed, either:

(a) a date within 28 days of award of contract, or
(b) such other date as may be agreed between the parties.

There is a great improvement on the vague wording of the Fifth edition which said only that the commencement date should be within a reasonable time after acceptance of tender. But it is still not entirely free from difficulty.

If there is no date in the appendix the contractor is entitled to assume that the commencement date will be within 28 days of the award of contract.

Any delay will be a breach of contract although the clause has nothing to say on recovery of the cost which could arise by the contractor having resources standing idle or in an extreme case in being thrown into winter working. The contractor could proceed with a common law claim for damages or perhaps for *quantum meruit*. In the old case of *Bush* v. *Whitehaven Port* (1888) the employer's delay in giving possession of the site in June as the contract specified threw the contractor into winter working. The court held the contractor was entitled to recovery on a *quantum meruit* basis or damages for his increased expenditure.

'Date as may be agreed'

However, what is the position if the engineer, seeking to avoid the rigidity of a specified date or a specified period, writes in the appendix not a date, but the phrase 'to be agreed'? Does this bring the provision in clause 41(1) (c) for a date to be agreed into play? The answer is probably not. The law does not favour agreements to agree and in reality one of the parties would have to accept the wish of the other.

The purpose of the 'to be agreed' provision is probably to give flexibility on the 28 day period so that, instead of being fixed with a commencement date no later than 28 days after the award of contract, the parties can by genuine agreement defer the date. This is a practical measure since there is nothing in the 28 day provision which requires the engineer to fix the commencement date with any consideration for the mobilisation needs of the contractor. In other words the engineer has 28 days in which he can theoretically say 'start tomorrow'! The wording of the clause might even be read to allow a date to be set retrospectively although that is clearly not intended.

Effect of the commencement date

One advantage of specifying the works commencement date in the appendix is that the formality is completed and cannot later be overlooked.

In the ICE Conditions there is a time for completion not a date for completion, thus the principal purpose of fixing the commencement date is to establish the completion date.

Without a firm completion date, the provisions for liquidated damages for delay do not apply.

There is undoubtedly an omission in clause 41(1)(c) in not requiring the agreed date to be fixed in writing. Words such as 'and confirmed in writing by the Engineer' should be added to avoid the problem in *Kemp* v. *Rose* (1858) where the date for commencement was omitted from a written contract and the court declined, in the face of conflicting oral evidence, to set a date.

Start of the works

Clause 41(2) requires the contractor to start the works on, or as soon as reasonably practicable after, the commencement date and thereafter to proceed with due expedition and without delay.

These are obligations the contractor must take seriously because, apart from any liability for liquidated damages for late completion, there are provisions in clause 63 for determination of the contractor's employment for failing without reasonable excuse to commence the works in accordance with clause 41 or failing to proceed with due diligence.

Reasonably practicable

The question of what is a 'reasonably practicable' time in clause 41(2) is a matter of fact to be determined in the light of all circumstances. Factors to be taken into account could include:

(a) the notice given by the engineer for commencement
(b) the complexity of the works and the need for pre-planning
(c) delivery times on plant and materials
(d) the contractor's programme
(e) the time for completion.

'Due expedition'

It is not clear why the phrase 'due expedition' is used in clause 41 and 'due diligence' in clause 63; or whether they are meant to be the same thing. In everyday language, expedition implies speed and promptness whereas diligence can imply speed, promptness or caution.

Apart from the doubtful linkage to clause 63 there is no express sanction in the contract for failing to proceed with due expedition.

The engineer has warning powers under clause 46 when he considers the contractor's progress too slow to complete on time and the contractor is then required 'to expedite' progress to complete on time. Commentators' views are mixed as to whether this is a sanction or mere verbiage. See the comment on clause 46 later in this chapter.

'In accordance with the contract'

Commentators have also puzzled over the final phrase in clause 41(2) 'in accordance with the Contract'. It probably means no more than the contractor shall proceed in accordance with his general obligation to complete on time. However, if the phrase is to be attached to 'expedition' and 'delay' its meaning is perhaps to emphasise the provisions of clause 46.

10.3 Clause 42 – possession and access

The main improvement in clause 42 is in presentation for the single paragraph clause of the Fifth edition is now divided into ten sub-clauses. This undoubtedly makes for easier reading and understanding.

The clause makes it very clear that the contractor is not entitled to full possession on commencement although the exact nature of the employer's obligation is still not as clearly expressed as it might be – see below.

Prescriptions

Clause 42(1) states that the contract may prescribe:

(a) the extent of portions, and
(b) the order of portions

of which the contractor is to be given possession from time to time. It further states that the contract may prescribe:

(a) the availability and nature of access to be provided by the employer, and
(b) the order in which the works shall be constructed.

The provisions on access are new to the Sixth edition but the others were to be found in clause 42 of the Fifth but with not the same contractual effect as here. That is because of the new provision in clause 14 requiring the contractor to have regard to the clause 42 prescriptions in his programme. The point of most concern relates to the order of construction.

Order of construction

There are clear dangers in fixing the order of construction by pre-contract requirements because any directions on how the contractor is to operate can rebound on the employer as claims if things go wrong. The employer starts only with the basic obligation not to impede the contractor in the performance of his obligation, which is to complete the works on time. But by prescriptions, restrictions, directions, orders or whatever, the employer can end up with a string of subsidiary obligations.

The case of *Yorkshire Water Authority* v. *Sir Alfred McAlpine & Son (Northern) Ltd* (1985), mentioned in chapter 6, illustrates the problem. In that case the contractor had submitted with his tender, as instructed, a bar chart and method statement, showing he had taken note of certain specified phasing requirements providing for the construction of the works upstream. The formal contract agreement incorporated the tender and the minutes of the meeting at which the method statement was approved.

The contractor maintained that in the event it was impossible to work upstream and after some delay work proceeded downstream. The contractor then sought a variation order under clause 51(1). The court held:

(a) the tender programme/method statement was not the clause 14 programme
(b) the incorporation of the method statement into the contract imposed an obligation on the contractor to follow it so far as it was legal or physically possible to do so
(c) the method statement, therefore, became a specified method of construction and the contractor was entitled to a variation order and payment accordingly.

Although the *Yorkshire Water* case related to a tender method statement it is suggested that, had the case been under Sixth edition conditions, the employer could have fallen into exactly the same trap under clause 42 prescriptions. The contract would have prescribed the order and the clause 14 programme would have to follow it.

Provision of site and access

Clause 42(2) states the employer's obligations in giving possession of the site and access thereto.

Under clause 42(2)(a) the employer is to give possession and access on the commencement date of so much of the site as is required to enable the contractor to commence and proceed. This is made 'Subject to sub-clause (1)'. This can be read as imposing a dual obligation on the employer but it is probably intended simply as a statement of the restrictions on the contractor's rights.

The contractor's programme

Under clause 42(2)(b) the employer is to give possession during the course of the works as required by the contractor's clause 14 programme and such further access as necessary to enable the contractor to proceed with due despatch.

The linkeage between possession and the contractor's clause 14 programme can be criticised in that it places contractual obligations on the employer arising out of a post-contract document. However the Sixth edition does at least refer to 'the programme which the Engineer has accepted' whereas the Fifth edition said only 'the programme referred to in Clause 14'.

Since the reference to the prescriptions in sub-clause (1) is omitted from clause 42(2)(b) it must be assumed that the engineer has accepted a clause 14 programme with those prescriptions. What is not clear is why the obligation on 'further access' is described in terms of 'due despatch' and not the clause 14 programme. It would be odd if the contractor, having got ahead of programme, could demand further access but not further possession.

In connection with this last point it may be worth emphasising two sides of the programme/possession issue:

(a) the engineer binds the employer to give possession to the clause 14 programme he accepts, and
(b) the contractor is bound by his clause 14 programme on

possession and must suffer the consequences of departing from it.

Failure to give possession

Clause 42(3) provides for the contractor to obtain an extension of time and recovery of additional cost for any delay or cost caused by the employer's default in giving possession.

The wording of the clause is different from that in other claim clauses and from that in the Fifth edition. The usual phrase for extension is 'the Engineer shall take such delay into account in determining any extension of time to which the Contractor is entitled'. Here it is, 'The Engineer shall determine any extension of time to which the Contractor is entitled'. There may be nothing in this or it may be intended that the engineer has a duty to deal with this extension without an application from the contractor.

On cost also the wording is more directory than elsewhere in the Conditions. Then there is the unusual and new closing sentence, 'The Engineer shall notify the Contractor accordingly with a copy to the Employer'. Notify the contractor of what? This can only be the extension of time and cost he considers the contractor entitled to. Again a duty on the engineer not to be found elsewhere in a claim clause.

It is difficult to avoid the conclusion that a breach by the employer under clause 42 is to be treated differently than other breaches. There may be some logic in this because there is a line of thought which says that failure to give possession is generally a 'fundamental' breach entitling the contractor to determine. With such determination not provided for in these Conditions it may be instead that the burden is put on the engineer to ensure that the contractor is compensated for breach.

Failure to give access

The Sixth edition has followed the Fifth edition in saying nothing about the contractor's rights if there is failure by the employer to give access as specified. But then the Fifth edition had no counterpart to clause 42(1) (c) by which the contract can prescribe the availability and access to be provided by the employer.

This may be a simple omission on the part of the draftsmen or it may go to the deeper point that while the employer is bound to provide possession of the site he is not generally bound to provide access. Nevertheless when the contract does say that access will be provided by the employer, failure to provide such access must be a breach for which the employer is liable in damages.

Access provided by the contractor

Clause 42(4) states that the contractor shall bear the costs of any access additional to those provided by the employer and shall bear the costs of any additional facilities outside the site.

The wording is slightly different to that in the Fifth edition which referred only to 'special or temporary wayleaves' in connection with access and 'additional accommodation' outside the site.

10.4 Clause 43 – time for completion

Clause 43 provides formal confirmation of the contractor's obligation to complete the works within the time stated in the appendix.

In the form of tender the contractor gives an undertaking to so complete but the form of agreement refers only to construction in conformity with the provisions of the contract.

Sectional completion

The obligation of completion in clause 43 extends to any section for which a particular time is stated in the appendix. The scheme in the Sixth edition is that there should be stated in the appendix either:

(a) a time for completion of the whole of the works, or
(b) times for various sections with the 'remainder' of the works forming a final section.

This is a much different scheme to that in the Fifth edition which incorporated sections into the whole. The change has been made to facilitate the calculation of liquidated damages for sections.

Note that all times for completion of sections run from the single works commencement date and are not intermediate periods.

Contractor's own times

The Sixth edition now makes provision for the practice, long used in the building industry and gaining popularity in civil engineering, of allowing the contractor to state his own times for completion. It does this by its two part appendix to the form of tender. The engineer/employer can either insert required times for completion in part 1 or allow the contractor to insert his own times in part 2.

The advantages of giving the contractor freedom are twofold. Firstly, tenders can be compared on time as well as price and secondly, disputes on shortened programmmes should largely be avoided.

Completion of phases (or parts)

Clause 43 places no obligation on the contractor to complete phases or parts of the works by times which may be specified in the specification, bills or drawings, unless those phases or parts are designated as sections and are properly included in the appendix.

Correspondingly there is no obligation on the employer to take possession of a phase or part not so designated unless it has been occupied or put into use. By clause 48 the employer is however obliged to accept any section for which the engineer has issued a certificate of substantial completion.

Damages for late completion of phases

Because the Conditions do not oblige the contractor to complete phases or parts, not designated as sections, within specified times, the provisions for liquidated damages do not apply, and cannot be applied, to late completion of phases or parts.

In *Bruno Zornow (Builders) Ltd* v. *Beechcroft Developments Ltd* (1989) a contract was negotiated for a housing development on the basis of a first tier tender which showed a detailed programme to complete in 16 months and second stage agreements for completion of the work in two overlapping phases. The architect calculated liquidated damages of £40,000 based on the stipulated rate of £200 per week per block from the date shown on the original works programme and the contractor sued for the return of this amount. It was held by Judge Davies QC:

'(i) the contract did not incorporate documents which specified dates for sectional completion but only phased provisions for the transfer of possession;
(ii) a claim for liquidated damages could only be made in respect of failure to meet specified completion dates and not failure to meet transfer of possession dates – which operated on a consent basis;
(iii) no term would be implied for any sectional dates for completion.'

A similar situation arose in *Turner v. Mathind* (1986) where there was a clear requirement in the bills for phased completion but the sectional completion supplement was not used and the appendix contained only a rate for liquidated damages for late completion of the whole of the works for £1000 per week. All attempts by the employer to justify deduction of liquidated damages for failure by the contractor to meet the phasing dates failed. It was not appropriate that the employer should either pro-rata the stipulated damages to the number of phases or apply the stipulated rate to each phase.

This was the case in which both Lord Justice Parker and Lord Justice Bingham made some intriguing comment, albeit *obiter*, and not therefore binding authority, on the apparent lack of remedy for late completion of phases and suggested there could be a case for ordinary as opposed to liquidated damages.

Completion within the time allowed

Both clause 43 and the appendix state the contractor's obligation to be to complete 'within' the time allowed. The contractor is therefore entitled to finish early. See the comment in section 6.8 on clause 14 on this.

10.5 Clause 44 – extension of time for completion

Clause 44 defines the events which entitle the contractor to an extension of time and sets out the procedures and rules for the contractor and the engineer to follow.

The major change from the Fifth edition is the introduction of a new procedure termed 'assessment of delay' which it is hoped will eliminate argument on whether the contractor is entitled to an extension of time when he can still complete within the time allowed.

The layout of the clause has been greatly improved and, in the process, some of the wording difficulties of the Fifth edition have been eliminated.

Other changes which should be noted are that time limits have been introduced for making assessments and, in particular, that interim extensions should now be granted 'forthwith'.

Purposes of extension provisions

A contractor is under a strict duty to complete on time except to the extent that he is prevented from doing so by the employer or is given relief by the express provisions of the contract.

The effect of extending time is to maintain the contractor's obligation to complete within a defined time and failure by the contractor to do so leaves him liable to damages, either liquidated or general, according to the terms of the contract. In the absence of extension provisions, time is put at large by prevention and the contractor's obligation is to complete within a reasonable time. The contractor's liability can then only be for general damages but first it must be proved that he has failed to complete within a reasonable time.

Extension of time clauses, therefore, have various purposes:

(a) to retain a defined time for completion
(b) to preserve the employer's right to liquidated damages against acts of prevention
(c) to give the contractor relief from his strict duty to complete on time in respect of delays caused by designated neutral events.

Contrary to common belief, extension of time clauses are not solely for the benefit of the contractor; nor are they an essential part of the financial claims procedure. In fact there is no linkage in the ICE Conditions between the award of an extension of time and recovery of extra cost although in practice it is well known that the maxim 'get time first and the money will follow' is effective.

Relevant events

Clause 44(1) defines the events which entitle the contractor to an extension of time. They are the same as in the Fifth edition and fall into five categories:

(a) variations under clause 51(1)
(b) increased quantities under clause 51(4)
(c) causes of delay referred to in the Conditions
(d) exceptional adverse weather
(e) other special circumstances.

Delays caused by nominated sub-contractors are not relevant events although in the Fifth edition, but not in the Sixth, delays caused by the operation of the forfeiture provisions on nominated sub-contractors were mentioned in clause 59.

Increased quantities under clause 51(4) do not necessarily add to the work content of the contract – it may simply be that certain work shown on the drawings, and of necessity in the contract, has been omitted or understated in the bills of quantities.

Entitlement to an extension then becomes a matter of judgment on what is fair.

Delays referred to in the Conditions

The delays referred to in the Conditions are most obviously those in clauses with specific reference to clause 44 but there can be delays under other clauses without such reference – e.g. clause 32 (fossils); clause 20 (excepted risks) – and here unless an instruction under clause 13 or a variation under clause 51 is involved, the extension would have to be given under any special circumstances. Clauses which do refer to clause 44 include:

clause 7(4) – late drawings and instructions
clause 12(2) – adverse physical conditions or artificial obstructions
clause 13(3) – instructions causing delay
clause 14(8) – delay in engineer's consent to the contractor's methods or because of the engineer's requirements
clause 27(6) – variations relating to public utilities
clause 31(2) – facilities for other contractors
clause 40(1) – suspension of work
clause 42(3) – failure to give possession.

Adverse weather

Adverse weather of itself does not give any grounds for non-performance of contractual obligations. Unless there are provisions in the contract offering relief, the contractor is deemed to have taken all risks from weather. However, most standard forms of construction contract make some provision for extension of time in respect of exceptional adverse weather.

The extent to which adverse weather applies as a relevant event depends on the wording of the contract. The ICE Conditions use the phrase 'exceptional adverse weather' conditions but it is not just the phrase which has to be considered but also its context; there has to be delay, not just exceptionally adverse weather. This may seem rather theoretical but it has become so common for contractors to apply for extensions of time on the grounds that the weather has been worse than average that sight can become lost of the need for proof of delay. The practice of obtaining local weather records and comparing them on a year to year basis, or on a particular year against average, may show that the weather has been exceptional but it says nothing about delay.

The point came up in the case of *Walter Lawrence and Son Ltd* v. *Commercial Union Properties (UK) Ltd* (1984) where a contractor was suing for return of amounts deducted as liquidated damages. It was held that:

> '...When considering an extension of time under clause 23(b) of JCT 63, on the ground of 'exceptionally inclement weather' the correct test for the architect to apply is whether the weather itself was 'exceptionally inclement' so as to give rise to delay and not whether the amount of time lost by the inclement weather was exceptional...'

Another matter of interest arose in the *Walter Lawrence* case in respect of the time at which the weather should be assessed. The architect in correspondence had said: 'It is our view that we can only take into account weather conditions prevailing when the works were programmed to be put in hand, not when the works were actually carried out.'

The contractor refuted this and claimed that his progress relative to programme was not relevant to his entitlement to an extension. It was held that the effect of the exceptionally inclement weather is to be assessed at the time when the works are actually carried out and not when they were programmed to be carried out even if the contractor is in delay.

It should be added however that the judge in the *Walter Lawrence* case drew a distinction between delays which occur during the original or extended time for completion and delays after the due date when the contractor is in culpable delay.

'Other special circumstances'

There are differing views on what can be included within 'other special circumstances'.

On one view the wording of clause 44(1)(e) is so wide – 'other special circumstances of any kind whatsoever which may occur' – that there are no exclusions. The other view is that catch-all phrases cannot include for unspecified breaches of contract by the employer. In *Fernbrook Trading Company Ltd* v. *Taggart* (1979) it was held that the words 'any special circumstances of any kind whatsoever' were not wide enough to empower the engineer to extend time for delays caused by the employer's breaches of contract in making late payments on interim certificates.

The point at issue in seeking to exclude unspecified breaches by the employer is that, if he has caused delay and there is no provision for extension, then the liquidated damages provisions fall and the employer can only recover such general damages as he can prove for the contractor's failure to complete within a reasonable time.

Quite apart from the above debate there are no fixed rules on what can, or should, be included within special circumstances. Matters contemplated by the contract but not cross-referenced to clause 44 should be non-contentious, but for strikes, insurance-covered delays and the like there is no consistency of treatment by engineers or by arbitrators and very little to assist by way of guidance.

Contractor's application

Clause 44(1)(a) provides that, should the contractor consider any of the relevant events entitles him to an extension of time, he shall:

> 'within 28 days after the cause of delay has arisen or as soon thereafter as is reasonable'

deliver to the engineer full and detailed particulars in justification of the extension claimed.

Unlike the position in many standard forms of construction, the contractor is not required to give notice of delay unless he is seeking an extension.

Although the wording of clause 44(1) is far more precise than that of the Fifth edition where the reference to delay is in the wrong place, it is still open for the contractor to claim for an extension of time, whether or not he expects to overrun the specified time for completion, and indeed to ensure that the assessment of delay procedure in clause 44(2) is activated, the contractor would be wise to do so.

Failure by the contractor to make an application under clause 44(1) does not automatically disentitle him to an extension. The engineer is required to assess the contractor's entitlement at the date for completion and on the issue of the certificate of substantial completion whether or not the contractor makes an application. The power to do this is essential to preserve the employer's right to liquidated damages.

Where the contractor will be at risk is in making a late application, or no application at all, in respect of neutral events such as adverse weather. He will similarly be at risk in gaining his full entitlement for all events where, by his conduct, he has either prevented the engineer from investigating the delay or prevented it from being properly assessed. Moreover the contractor cannot complain that he has not been given an interim extension if he fails to make application.

Engineer's assessment of delay

Clause 44(2) is a completely new provision requiring the engineer to assess 'delay' on receipt of an application for an extension.

The clause requires the engineer to consider all the circumstances known to him at the time; to assess the delay; and to notify the contractor in writing of his findings. The engineer may also, in the absence of a claim from the contractor, make an assessment in respect of any delay he considers may have been caused by any of the events listed in clause 44(1).

The new provision has been introduced with two intentions:

(a) to establish and place on record details of delays for availability in subsequent disputes; and
(b) to avoid the granting of extensions of time when they are not strictly necessary in relation to the time for completion.

The first objective would have been improved by a requirement for the contractor to give notice of any delay, whether or not it involves extension of time; the second objective, by a definitive statement, that any extension of time would apply only to delay beyond the date for completion.

Clause 44(2)(b) only allows the engineer to take account of the circumstances defined as relevant events for extensions, so he is not entitled, although it might be useful, to issue a notice under this clause where it is apparent that the contractor has been delayed by circumstances within his control which give no entitlement to an extension.

An odd aspect of the assessment of delay provisions in clause 44(2), insofar that the provisions are supposed to eliminate the granting of extensions when completion can still be achieved within the original time, is that the contractor's application under clause 44(1) remains firmly stated as an application for an extension of time. It would have been better stated as either an application for assessment of delay or an extension of time if that is its purpose.

Interim extension of time

Having made an assessment of delay under clause 44(2) the engineer must then consider whether the delay fairly entitles the contractor to an extension of time.

Clause 44(3) provides that the engineer must either:

(a) grant any extension which is due forthwith in writing, or
(b) where the contractor has made a claim and no extension is considered due – inform the contractor without delay.

Fairly entitled

It is interesting that under clause 44(1) the contractor applies for extensions to which he is 'entitled' but under clause 44(3) the engineer grants only extensions to which the contractor is 'fairly entitled'. The 'fairly' test applies again in clause 44(4) but not in clause 44(5).

Granted 'forthwith'

The requirement that an interim extension should be granted 'forthwith' should be taken seriously by all engineers who wish to avoid claims from contractors for constructive acceleration. More is said on this under clause 46 later in this chapter but the very purpose of introducing interim extension procedures into the Fifth edition in 1973 was to ensure, as far as possible, that contractors got extensions of time in time to use them.

A minor point of disappointment is that the assessment of delay itself need not be made 'forthwith' so there is still a defence, albeit weak, for procrastination. Another minor lapse in the drafting is that there is no requirement in clause 44(3) for rejection of an application for an extension to be made in writing.

Assessment at due date

Clause 44(4) places a firm duty on the engineer to review the contractor's entitlement to extension of time not later than 14 days after the due date or extended date for completion of the works or any section. This must be done whether or not the contractor has made an application.

The engineer is to consider all the circumstances known to him at the time – that is to say he is not confined to those matters notified to him by the contractor. See under clause 44(5) below for comment on reductions.

The procedure under clause 44(4) is clearly intended to operate on a repeat basis and it must be put into effect after each extended completion date has expired.

Note that, if the engineer does not consider the contractor entitled to an extension of time, he is to notify both the employer and the contractor. This is presumably so that the employer can consider deducting liquidated damages under clause 47 although the reference in that clause to clause 44 certificates has been deleted from the Sixth edition as a condition precedent to deduction.

Final determination of extension

Clause 44(5) is the final part of the four stage extension procedure and it requires the engineer to make a final review of the contractor's entitlement within 14 days of the issue of the certificate of substantial completion of the works or any section.

The engineer is to certify to the contractor, with a copy to the employer the extension he considers due.

Reduction of extensions granted

The clause states that the final review shall not result in any decrease in any extension granted at the clause 44(3) and 44(4) assessments; that is at the interim or due date stages. This provision inevitably raises the question, can the earlier assessments under 44(3) and 44(4) produce reductions? Commentators are divided on this but logically, if an extension has been granted for an addition variation under 44(3) and there is subsequently a significant omission variation, it should be possible to recognise this at a later clause 44(3) assessment or at a clause 44(4) assessment.

Final review

Although clause 44(5) describes the engineer's review after the issue of certificate of substantial completion as his 'final review' it may not be his last review. If either the contractor or employer is dissatisfied with the extension granted either can call for an engineer's decision under clause 66 which effectively obliges the engineer to carry out a further review.

Sections

The provisions of clause 44 apply with equal effect to each and every section as to the whole of the works. Therefore each stage of procedure must be followed for each and every section. This is slightly different from the Fifth edition which did not expressly provide for assessments at due dates for completion of sections.

Time limits

It is unlikely that failure by the engineer to act within the 14 day time limits prescribed in clauses 44(4) and 44(5) would set time at large. It is probable that they must be regarded as 'directory' only following the decision on a similar point in a building contract in the case of *Temloc Ltd* v. *Errill Properties Ltd* (1987).

10.6 Clause 45 – night and Sunday work

Clause 45 remains identical to clause 45 of the Fifth edition.

It provides that none of the work shall be carried out at night or on Sundays without the written permission of the engineer unless unavoidable or necessary for safety reasons. Two other exceptions are given:

(a) any provision to the contrary in the contract, and
(b) work it is customary to carry out outside normal working hours or by shifts.

A further overriding exception is found in clause 46(2).

It is not clear whether the considerations of safety qualify 'unavoidable' as well as 'absolutely necessary' or whether unavoidable work can be carried out whatever its nature. As to work which is pre-planned for night or Sunday work, that would normally be expressed in the drawings or specification and would come within 'any provision to the contrary'.

10.7 Clause 46 – rate of progress and acceleration

The purpose of clause 46 is a mystery but it has all the makings of a nasty trap for the engineer. In the Sixth edition the clause has a new provision for acceleration and appropriately the purpose of this is equally obscure.

Rate of progress

Clause 46(1) places a duty on the engineer to notify the contractor in writing when at any time, in his opinion, the rate of progress is too slow to ensure substantial completion by the due date. The contractor is then obliged to take such steps as are necessary – and to which the engineer may consent – so as to complete by the due date. The provisions apply equally to sections and the whole of the works.

The clause commences with the words 'If for any reason which does not entitle the Contractor to an extension of time' and ends with the words 'The Contractor shall not be entitled to any additional payment for taking such steps'.

The first objection to these provisions is that the engineer is obliged to constantly monitor progress and issue notices, although the frequency of such notices is not stated, and failure to do so is a breach for which the employer is technically liable. The second objection is that there is no sanction on the contractor if he ignores the notices, other than in extreme cases of using the determination provisions of clause 63. In practice the contractor who is running late will make a commercial choice between incurring extra production costs or paying liquidated damages. This clause indicates that he has no such choice.

Constructive acceleration

The trap for the engineer, which can be costly to the employer, lies in the opening and closing words of the clause. What is the position if, after giving notice to the contractor that he must take steps to expedite progress, the engineer or an arbitrator in later review grants an extension of time? Can the contractor recover the costs of constructive acceleration?

There is no definitive authority on this but the position, it is suggested, is even stronger for the contractor than for constructive acceleration resulting from failure by the engineer to operate the provisions of clause 44 in granting extensions of time promptly.

As long ago as 1973 Sir William Harris in his defence of the Fifth edition said this:

> 'Not only may an employer's rights to liquidated damages be endangered if the extension of time provisions are not operated properly, but also he may find himself exposed to a claim for accelerated costs if the Contractor is pressed to complete by the original contract date and eventually proves his entitlement to an extension.'

In both *Fernbrook Trading Co. Ltd* v. *Taggart* (1979) and *Perini Corporation* v. *Commonwealth of Australia* (1969) on civil engineering contracts, the judgments support the proposition that there is breach of contract if the engineer fails to grant extensions in reasonable time.

Permission to work night or on Sunday

Clause 46(2) links back to clause 45 so that, if the engineer has served a clause 46 notice, he cannot then unreasonably refuse the contractor permission to work at night or on Sundays. The phase 'on site' has been added to clarify the need for permission.

Accelerated completion

The new clause 46(3) has nothing to do with constructive acceleration discussed above. It relates only to 'requested' acceleration where the employer or the engineer requests the contractor to complete within a time less than that in the appendix or as extended.

The clause does no more than provide that, if the contractor agrees to accelerate, any special terms and conditions of payment shall be agreed between the parties before action is taken. Clearly this cannot be binding on the parties if they choose to ignore it but it does have some practical effect in that the contractor can ask whatever price he likes.

This latter point offers a possible explanation for the inclusion of the new provision. There is a view relevant to the Fifth edition that the powers of the engineer under clauses 13 and 51 are so great that he can order acceleration and evaluate its cost under clause 52. If that possibility is in any way correct, then its operation is now clearly defeated by clause 46(3) in the Sixth edition.

Liquidated damages

Note that the added reference to clause 46(3) in clause 47(1) indicates that agreed acceleration could involve the payment of liquidated damages from an earlier date than would otherwise apply, but again this is clearly a point for agreement.

Chapter 11

Liquidated damages for delay

11.1 Introduction

This chapter examines clause 47 which provides for liquidated damages as the employer's remedy for late completion by the contractor.

Significant and highly visible changes have been made from the Fifth edition in respect of liquidated damages for sections and for the intervention of variations etc. after the due date for completion. Less visible, but equally significant, is the deletion of an engineer's certificate as a condition precedent to the deduction of liquidated damages. Various other changes of detail have been made.

11.2 Liquidated damages – generally

'The essence of liquidated damages is a genuine covenanted pre-estimate of loss.' So said Lord Dunedin in *Dunlop Tyre* v. *New Garage* (1915). The characteristic of liquidated damages is that loss need not be proved.

Most standard forms of construction contract are drafted to permit the parties to fix in advance the damages payable for late completion. When these damages are a genuine pre-estimate of the loss likely to be suffered, or a lesser sum, they can rightly be termed liquidated damages. In short, liquidated damages are fixed in advance of the breach whereas general, or unliquidated damages, are assessed after the breach.

Reasons for use

There are sound commercial reasons for using liquidated damages whenever possible. Firstly, because of the certainty they bring to the consequences of breach; and secondly because they avoid the expense and dispute involved in proving loss.

It is a mistake to assume that liquidated damages provisions are solely for the benefit of the employer. Liquidated damages provisions are beneficial to contractors for they not only limit the contractor's liability for late completion to the sums stipulated, but they also indicate to the contractor at the time of his tender the extent of his risk. Thus, if a contractor believes that he cannot complete within the time allowed he can always build into his tender price his estimated liability for liquidated damages. All that the employer gets out of liquidated damages is relief from the burden of proving his loss and usually, in construction contracts, the right to deduct liquidated damages from sums due to the contractor. To the extent that the employer's true losses may be greater than the stipulated level of liquidated damages he is disadvantaged by agreeing to a restrictive remedy.

Exhaustive remedy

One aspect of liquidated damages which needs to be emphasised is that if the provisions are valid they provide an exhaustive and exclusive remedy for the specified breach. This was one of the points confirmed in the case of *Temloc Ltd* v. *Errill Properties Ltd* (1987) where Lord Justice Nourse said:

> 'I think it clear, both as a matter of construction and as one of common sense, that if ... the parties complete the relevant parts of the appendix ... then that constitutes an exhaustive agreement as to the damages which are ... payable by the contractor in the event of his failure to complete the works on time.'

The effect of this is that the employer cannot choose to ignore the liquidated damages provisions and sue for general damages, nor can he recover any other damages for late completion beyond those specified. This is the point in chapter 10 on phased, as opposed to sectional, completion.

If the provisions fail

The rule is well settled that when a liquidated damages clause fails to operate because it is successfully challenged as a penalty, or fails because of some defect in legal construction, act of prevention or other obstacle, then general damages can be sought as a substitute. Lord Justice Phillimore in *Peak* v. *McKinney* (1970) said:

> 'If the employer is in any way responsible for the failure to achieve the completion date, he can recover no liquidated damages at all and is left to prove such general damages as he may have suffered.'

There is no firm legal ruling in English law that liquidated damages invariably act as a limit on any general damages which may be awarded as a substitute and the courts take a cautious approach to the matter. In *Rapid Building* v. *Ealing Family Housing* (1984) where the liquidated damages clause was held to have failed, neither Lord Justice Stephenson nor Lord Justice Lloyd would be drawn on the proposition that the quantum of general damages was limited to the quantum of liquidated damages.

Liquidated damages and penalties

The association between liquidated damages and penalties lies in the nature of the remedy – an agreed price to be paid for breach or non-performance. The parties may agree any price they wish; they are not bound by any rules and if the price they agree is clearly intended to penalise the defaulting party rather than to compensate and restore the position of the innocent party, that is a matter for the parties. The question which is of prime importance to the parties in agreeing their price is, will the courts assist them in enforcing payment?

Many legal systems do allow the recovery of penalties and it is something of a peculiarity of English law that the courts will look at the price irrespective of whether it is called liquidated damages or a penalty, and, if it is found to be a penalty, will limit damages to the amount flowing from the breach.

The background is that a penalty was originally held to be in the nature of a threat held over the other party *in terrorem*. The courts of equity took the view that, since a penalty was designed to secure performance, the promisee was sufficiently compensated by being indemnified for his actual loss and he was not entitled to demand a sum which, although fixed by agreement, might be disproportionate to the actual loss suffered. Common law now applies the same approach.

The legal effect of this is that when there is a penalty clause the plaintiff may sue either on the penalty clause, in which case he cannot recover more than the stipulated sum, or he can ignore the penalty clause and sue for breach of contract to recover damages in full. In either case damages can only be recovered to the extent they are proved. The original *in terrorem* aspect of penalties has been gradually diluted by the courts as they have dealt with disputes on whether stipulated sums entered in contracts to cover breaches are in law penalties or are liquidated damages. The rules on penalties are now applied to any stipulated sums which can be proved not to accord with a genuine pre-estimate of loss, even where this has clearly arisen unintentionally.

11.3 Clause 47(1) – liquidated damages for the whole of the works

Clause 47(1) (a) provides for liquidated damages where the whole of the works is not divided into sections. It says there 'shall' be included in the appendix a sum which represents the employer's genuine pre-estimate of damage likely to be suffered if the whole of the works is not substantially completed within the time allowed or as extended.

Two points are worth noting. One, that liquidated damages are mandatory under the Sixth edition. See also clause 47(4) (b) below. The Fifth edition left some scope for their omission. Two, the provision in the Fifth edition for liquidated damages to be a lesser sum than the genuine pre-estimate has been omitted – but see clause 47(4) (a) below on possible limitation.

The new reference to clause 46(3) indicates that the terms agreed for acceleration may include advancing the date on which damages become due.

Per week or per day

Clause 47(1) (a) and the appendix both allow for liquidated damages to be expressed per week or per day.

It is usually better to use 'per day' since damages 'per week' cannot be proportioned down for a part week. Thus the contractor is only liable for each full week of delay if weeks are used.

It is not necessary for liquidated damages to be expressed in the same time periods as the time for completion.

Contractor's liability

Clause 47(1)(b) states the contractor's liability to pay liquidated damages if he fails to complete on time. Clause 47(5) details the methods of payment.

Completion of parts

The final paragraph of clause 47(1) is a provision for proportioning down the specified figure for liquidated damages when parts of the works have been completed prior to the whole.

The general position is that unless the contract provides a mechanism for the issue of partial completion certificates and a corresponding mechanism for proportioning down liquidated damages, the contractor remains liable for full damages up to the date of total completion as in *BFI* v. *DCB* (1987) where the employer avoided loss by taking partial possession but still obtained full damages.

However, where a contract does make provision for the issue of partial completion certificates, albeit on a consent basis, then it is essential to have corresponding provisions for proportioning down liquidated damages. The courts will not imply such a term if it is missing and the liquidated damages clause will then fall for uncertainty or as a penalty.

Thus, in *Stanor Electric* v. *Mansell* (1988) when there was a single sum for damages for two houses and one was completed late, the absence of any contractual machinery for proportioning down the damages led to them being declared penalties. Consequently, most standard forms of construction contracts have proportioning-down clauses.

11.4 Clause 47(2) – liquidated damages for sections

Clause 47(2) repeats the provisions of clause 47(1) so that they apply when the whole of the works is divided into sections.

What happens is that the provisions in clause 47(1) for the whole of the works as a unity are effectively omitted and each section is given its own time and liquidated damages with the 'remainder of the works' as a further section. This avoids the complex and sometimes unworkable provisions of the Fifth edition which attempted to embrace damages for the whole of the works with damages for sections.

Clause 47(2)(c) confirms what is fundamental to the scheme, namely that liquidated damages in respect of two or more sections may run concurrently.

11.5 Clause 47(3) – damages not a penalty

Clause 47(3) states that sums paid by the contractor pursuant to clause 47 are paid as liquidated damages and not as a penalty. This provision is of no legal effect.

The terminology itself is not decisive and if the courts find, as a matter of construction, that liquidated damages are penalties or *vice versa* they will award accordingly. Thus, Lord Dunedin in *Dunlop Pneumatic Tyre Co. v. New Garage and Motor Co. Ltd* (1915) said:

> 'Though the parties to a contract who use the words "penalty" or "liquidated damages" may *prima facie* be supposed to mean what they say, yet the expression is not conclusive.
>
> The court must find out whether the payment stipulated is in truth a penalty or liquidated damages.'

11.6 Clause 47(4) – limitation of damages

Clause 47(4) contains two significantly different provisions.

Limitation of liability

Clause 47(4)(a) relates to where limitation of liability for sums payable as liquidated damages has been written in the appendix. This is, of course, at the discretion of the employer in the first instance although it is not inconceivable that figures might be entered after negotiations with the contractor.

This provision seems to have taken the place of the 'lesser sum' in the Fifth edition.

The statement in clause 47(4)(a), if no limit is stated then 'liquidated damages without limit shall apply', means only that the contractor's liability is not stopped at a ceiling for prolonged delay.

'Nil' damages or omission

Clause 47(4) (b) provides that if no sums are stated for liquidated damages in the appendix or if 'Nil' is written, then damages are not payable.

What this means is that the liquidated damages provisions do not become inoperative because of a blank or a nil entry; they remain in force but nothing can be recovered. The employer cannot sue for general damages for late completion because he has exhausted his remedy for the contractor's breach.

This was one of the matters which came to be considered by the Court of Appeal in the case of *Temloc Ltd* v. *Erril Properties Ltd* (1987) where the entry in the appendix to a JCT 80 contract was stated as £nil liquidated damages.

The contract was finished late and the employer/developer, who was himself liable to the property purchaser for damages, sought to recover them as general damages from the contractor. The employer put up three arguments:

(a) that he had alternative rights to claim liquidated or general damages – that argument failed
(b) that the liquidated damages provisions were rendered inoperative by failure of his architect to observe the administrative requirements of the contract – that argument failed
(c) that 'nil' liquidated damages rendered inoperative the provisions – that argument also failed.

On this latter point Lord Justice Nourse said:

> 'I find it impossible to attribute to parties who complete the appendix in one way or the other an intention that the employer shall have the option of claiming damages of precisely the same character but in an unliquidated amount.'

Clause 47(4) (b) it may be noted goes further than the decision in the *Temloc* case by equating a blank entry with a nil entry. The effect of this is that, if the employer genuinely wants the right to general damages as opposed to liquidated damages for late completion, the whole of clause 47 needs to be deleted from the contract with corresponding amendments made to other clauses.

11.7 Clause 47(5) – recovery of damages

Clause 47(5) has two broad provisions:

(a) the employer may 'deduct' or require the contractor 'to pay' the amount due, and
(b) interest must be paid on sums reimbursed to the contractor for any overpayment.

Conditions precedent

It is noteworthy that clause 47(5)(a) entitling the employer to deduct or require payment comes into operation without any condition precedent other than that in clause 47(1)(b) 'if the Contractor fails to complete ... within the time'. This is a change from the Fifth edition which required an engineer's certificate and it should be a matter of some concern to contractors who may find damages have been deducted when the engineer is still assessing what may be an undisputed claim for an extension.

Option on deduction

Note also the wording 'The Employer may' at the start of clause 47(5). This indicates, it is suggested, more than simply an option to deduct or require payment but also a general option on whether to claim payment at all.

Repayment with interest

Clause 47(5) provides that should extensions of time be granted which reduce the contractor's liability for damages, then any amounts overpaid shall be returned with interest. Surprisingly, not all standard forms allow for this.

The rate of interest is to be that provided for in clause 60(7) – 2% above base rate.

11.8 Clause 47(6) – intervention of variations

Clause 47(6) deals in an unusual way with the unresolved question of what extension, if any, is due when delay for which the contractor is not responsible occurs after the due completion date.

Instead of attempting to resolve whether an extension is due up to the date at which the delay ends, or whether it is due only for the period of delay itself, this provision avoids the argument by suspending damages for the period of delay. In principle this is a good approach but the drafting of the clause is likely to lead to difficulties.

The first problem is that the grounds for suspending liquidated damages are too wide and it is the engineer's 'opinion', not a test of fairness, which applies. The stated grounds are:

(a) a variation order under clause 51
(b) adverse physical conditions or artificial obstructions within clause 12
(c) any other situation outside the contractor's control.

The last item goes well beyond the relevant events for extensions of time. Thus any delay by weather would seem to qualify, whether exceptionally adverse or not, and it does not seem to be open to the engineer to consider whether it is fair for suspension of damages to be made: he can only form an opinion on whether there has been delay.

Contractors in culpable delay and paying liquidated damages will readily find a range of matters outside their control to offer as excuses for continuing delay – not least, lack of performance by their sub-contractors or suppliers.

Beyond the contractor's control

The possibility of general phrases in extension provisions being used to cover sub-contractor delay arises from the interpretation of 'a cause beyond the contractor's control' in *Scott Lithgow* v. *Secretary of State for Defence* (1989). Although that case related to a right of payment the same decision might well have been given in relation to extension of time. The case does indicate the need for great caution in drafting general phrases which give rights and obligations.

'Part of the Works'

The second problem lies in the use of the phrase 'that part of the

Works' with regard to both delay and suspension of damages. It is not obvious why delay to part of the works, particularly if non-critical, should lead to suspension of damages.

Nor is it clear how the level of suspended damages should be calculated. Perhaps the intended meaning is that damages should be suspended if there is delay to a section or the whole of the works, but by definition in clause 1(1)(u) a section has its own identity whereas a part is undefined.

The final problem is interpretation of the last paragraph of the sub-clause. The wording throws into question the whole purpose of the suspension of damages provision. Clearly the suspension does not invalidate entitlement to damages before the period of delay, nor would there be liability for interest on such damages.

Chapter 12

Completion and defects

12.1 Introduction

This chapter covers the following clauses:

clause 48 – certificate of substantial completion
clause 49 – outstanding work
clause 50 – searching for defects
clause 61 – defects correction certificate
clause 62 – urgent repairs.

The changes from the Fifth edition relate mainly relate to terminology but a subtle alteration in clause 48 in the procedure for issue of a certificate of substantial completion where the employer has taken premature use of part of the works needs to be noted.

12.2 Meaning of completion

Disputes on whether completion, be it practical completion or substantial completion, has been achieved are commonplace. Contractors want early completion to reduce insurance and other liabilities and to give part release of retention monies; employers are less enthusiastic, fearing problems with unfinished works or simply not being ready for occupation. There is no single definition which will meet the facts of all cases and satisfy the tests of different contracts. Judicial comment is invariably restricted in its generality.

Completion

The word 'completion' implies the strict test of full completion of the actual work which has to be done. If used consistently in a contract without a qualifying adjective it is unlikely that the courts would imply that it could apply to work with items to finish or work with patent defects still to remedy.

However, the word 'completion' is rarely so used and in *Emson Eastern Ltd* v. *EME Development Ltd* (1991) the court had to decide whether the issue of a certificate of practical completion under a JCT 80 contract constituted 'completion of the works' as mentioned in the determination clause of the contract. Judge Newey QC held that it did. He said:

> 'In my opinion there is no room for "completion" as distinct from "practical completion". Because a building can seldom if ever be built precisely as required by drawings and specification, the contract realistically refers to "practical completion", and not "completion" but they mean the same.'

In any event the change in the Sixth edition from 'Certificate of Completion' in the Fifth edition back to the old phraseology of the Fourth edition 'Certificate of Substantial Completion' is to be welcomed. It was never possible to reconcile the provisions of clause 48 of the Fifth with a true certificate of completion.

Practical completion

Building contracts use the phrase 'practical completion'. Here a few guidelines can be drawn from the decisions of the courts. In *H.W. Nevill (Sunblest) Ltd* v. *William Press & Sons Ltd* (1981) a problem arose from defective groundworks in a preliminary works contract found after the issue of a certificate of practical completion. Judge Newey QC said:

> 'I think that the word 'practically' in clause 15(1) gave the architect a discretion to certify that William Press had fulfilled its obligation under clause 21(1), where very minor *de minimis* work had not been carried out, but that if there were any patent defects in what William Press had done the architect could not have given a certificate of practical completion.'

In *City of Westminster* v. *J. Jarvis & Sons Ltd* (1970), a building case concerning whether the contractor could get an extension of time for carrying out replacement of faulty piling undertaken by a nominated sub-contractor but not discovered until after that sub-contractor had been given a certificate of completion of his work, Viscount Dilhorne said:

> 'From these provisions there are, in my opinion, two conclusions to be drawn: first that the issue of the certificate of practical completion determines the date of completion, which may of course be before or after the date specified for that in the contract; and secondly, that the defects liability period is provided in order to enable defects not apparent at the date of practical completion to be remedied. If they had been then apparent, no such certificate would have been issued.
>
> It follows that a practical completion certificate can be issued when, owing to latent defects, the works do not fulfil the contract requirements; and that under the contract, works can be completed despite the presence of such defects. Completion under the contract is not postponed until defects which became apparent only after the work had been finished have been remedied.'

From these cases and others it is suggested that the following rules apply to the phrase 'practical completion' as used in building contracts:

(a) practical completion means the completion of all the construction work to be done
(b) the contract administrator may have discretion to certify practical completion where there are minor items of work to complete on a *de minimis* basis
(c) a certificate of practical completion cannot be issued if there are patent defects
(d) the works can be practically complete notwithstanding latent defects.

ICE Minor Works Conditions

It is worth noting that, unusually for a civil engineering contract, the ICE Minor Works Conditions use the phrase 'practical completion'. In those conditions however it is stated:

> 'Practical completion of the whole of the Works shall occur when the Works reach a state when notwithstanding any defect

or outstanding items therein they are taken or are fit to be taken into use or possession by the Employer'.

Substantial completion

The phrase 'substantial completion' is probably more flexible than 'practical completion' and the provisions for dealing with outstanding works in the ICE Conditions suggest that it is not the *de minimis* principle which applies to such works but whatever is acceptable to the engineer or is compatible with the occupation and use by the employer.

It can certainly be argued with some force that in both the Fifth and Sixth editions the test for substantial completion of the works or parts of them is that they are either in use, or fit to be put in use, by the employer.

12.3 *Clause 48 – certificate of substantial completion*

Clause 48(1) provides that when the contractor considers that the whole of the works or any section, as defined in the appendix, is substantially complete he may give notice in writing to the engineer or the engineer's representative. Such notice is to be accompanied by an undertaking to finish any outstanding work during the defects correction period.

Contractor to give notice

The burden of giving notice falls on the contractor but it is not expressed in the mandatory term of 'shall give notice' but in the permissive term of 'may give notice'.

Clause 48(4) does allow the engineer to act on his own initiative in issuing certificates in respect of parts of the works and possibly this power could be used by the engineer for the whole of the works if the contractor failed to give any notice.

The contractor is entitled to give notice when he considers the works have been substantially completed and passed any final test in the contract. He will form his own view on the meaning of substantial completion and what level of outstanding works and remedial works should be acceptable.

Contractor to apply

Note that strictly the contractor does not apply for a certificate of substantial completion under clause 48(1). He does so under clause 48(3) where there is premature use by the employer but under clause 48(1) the contractor merely gives notice that he considers the works to be substantially complete and undertakes to finish the outstanding work.

In the Fifth edition, in clause 48(1), it was stated that the contractor's notice was deemed to be an application for a certificate. That has been omitted from the Sixth edition.

Note also that the contractor's undertaking to finish outstanding work is 'in accordance with the provisions of Clause 49(1)' whereas in the Fifth edition it was during the maintenance period.

Notice to engineer's representative

Although the contractor is entitled to give his notice to the engineer's representative, no powers are conferred in clause 48 on the engineer's representative so he can do no more than pass the notice on to the engineer. He cannot even have delegated powers because this is prohibited under clause 2(4) (c).

Engineer's response to notice

Clause 48(2) requires the engineer to respond within 21 days of the 'date of delivery' of the contractor's notice by either:

(a) issuing a certificate of substantial completion, or
(b) giving instructions in writing to the contractor specifying the work to be done before the issue of the certificate.

Date of certificate

The certificate, which is to be issued to the contractor and copied to the employer, is to state the date on which the works were substantially completed in the opinion of the engineer.

The certificate will, therefore, carry two dates – the date of issue and the date of substantial completion. The latter will normally be a prior date but note this does not upset clause 20 on care of the works which operates from the issue date.

Instructions in writing

The wording of clause 48(2)(b) requiring the engineer to give 'instructions' on work to be done before the issue of a certificate is not ideal. The engineer should be very careful in giving any instructions which might give rise to claims under clause 13. Instructions telling the contractor how to do the work should certainly be avoided.

Even instructions doing no more than scheduling the work to be done could possibly give rise to a claim under clause 13 if the contractor could show that it had caused him to incur cost beyond that which could reasonably have been foreseen at the time of tender.

The contractor must, of course, comply with the instructions of the engineer and then pursue his remedies under the contract.

There is nothing in clause 48 itself to suggest that the engineer can relate his 'instructions' to an undertaking by the contractor to complete outstanding work within a specified time, as envisaged by clause 49(1). In practice, however, the two are likely to be related.

Completion of work to be done

The final sentence of clause 48(2) suggests that when the contractor has completed the specified works to the satisfaction of the engineer he is entitled to a certificate without further notice or application within 21 days of such completion. It is not unreasonable that the burden of monitoring the contractor's performance should fall on the engineer for it is after all, at this stage, no more than a matter of the engineer's opinion against the contractor's opinion.

Contractor's design

The engineer must be aware when considering his action under clause 48(2) that, somewhat unusually, one clause of the contract, clause 7(6)(b), overrides his opinion and by that clause he is prevented from issuing a certificate covering any part of the permanent works designed by the contractor until all drawings and manuals have been submitted and approved.

Engineer's failure to respond

Clause 48 is silent on the position which applies if the engineer fails to respond to the contractor's notice, in time or at all.

The contractor has a strong case for arguing that once the 21 days has passed it is too late for the engineer to give instructions on works to be completed; and he has an equally strong case for arguing that his notice should be effective in fixing the date of substantial completion.

Regrettably, the way clause 47 is now drafted, the contractor may unjustly suffer the deduction of liquidated damages until the matter is resolved.

Premature use by employer

Clause 48(3) provides for the issue of a certificate of substantial completion for any 'substantial' part of the works which has been occupied or used by the employer 'other than as provided in the Contract'.

The procedure is simply that the contractor shall request and the engineer shall issue. There is no reference here to 'the opinion of the Engineer' there is no need for the contractor to give an undertaking on outstanding work; nor is there any provision for the engineer to refuse until further work has been completed. These are significant changes from the Fifth edition.

There is, of course, still scope for dispute on what is meant by a 'substantial part of the Works' and 'other than as provided in the Contract'.

What is a substantial part is a matter of fact to be decided on the circumstances of the particular case. What the contract 'provides' has to be determined by reference to what is contained in the Conditions, specification, drawings, bill of quantities and any other contract documents.

It is not clear if the restriction for contractor's design in clause 7(6)(b) applies in the event of premature use by the employer and the probability is that it does not.

Part of the works

Clause 48(4) provides that the engineer may issue a certificate of substantial completion for any part of the works which has been substantially completed and has passed any final test.

This is worded as an entirely discretionary power but it is suggested that the engineer should not exercise this power without first consulting with the employer for it is his interests which are at stake.

It is somewhat odd that the provision operates without any notice from the contractor, or any undertaking on outstanding work (which is deemed to have been given) and it could apparently be used against the contractor's wishes. Interestingly the deemed undertaking relates to completion of the outstanding work during the defects correction period and not in accordance with the provisions of clause 49(1) as in clause 48(1).

Reinstatement of ground

Clause 48(5) states only that a certificate of substantial completion shall not be deemed to certify completion of surfaces requiring reinstatement unless the contract so provides.

12.4 Clause 49 – outstanding work and defects

Clause 49 covers the respective rights of the contractor and the employer during the defects correction period in respect of outstanding work and defects. The difficult wording of clause 49(2) of the Fifth edition has been avoided by splitting the contractor's obligations for outstanding works and for defects into clause 49(1) and 49(2) in the Sixth.

Outstanding work – clause 49(1)

The general obligation in both the Fifth and Sixth editions is that the contractor shall complete any outstanding work 'as soon as practicable' during the defects correction period. Clause 49(1) of the Sixth edition, however, has the useful additional provision that the contractor's undertaking, given under clause 48(1) may specify a time for completion.

Although the wording in clause 49(1) does not precede 'defects correction period' with 'relevant' as does clause 49(2), the obligation to finish outstanding work is clearly intended to apply to each certificate issued.

There is nothing in clause 48 to force the contractor into giving a timed undertaking, but perhaps it is thought that agreement on timing will assist the engineer in his consideration of whether or not to issue the required certificate.

'As soon as practicable'

It is certainly a matter of common contention that contractors are disposed to leave the execution of outstanding works to the end of the defects correction period rather than tackle them early on. Contractors argue, partly from commercial considerations, but also with some practical justification that it is not practicable to undertake such works on a piecemeal basis and that what is 'as soon as practicable' has to be a matter for their judgment not that of the engineer.

It is not fully clear whether clause 49(4) gives the employer a remedy if the contractor defaults in carrying out the outstanding works as soon as promised or as soon as practicable. See the comment below on the meaning of the phrase 'any such work' in clause 49(4). The general position would seem to be that failure by the contractor to comply would leave him liable to the employer for such damages as could be proved because this is not a matter of late completion of the works itself but a separate breach with the potential to disrupt the employer's occupation and use of the works.

Costs of repair

Clause 49(3) confirms the obvious point that works of repair etc. carried out under clause 49(2) are to be at the contractor's cost to the extent they are due to his neglects or defaults, and otherwise at the employer's expense.

The clause provides that the value of work for which the contractor is to be paid shall be ascertained and valued as if it were additional work. That can be either as a variation or as daywork.

Work of repair – clause 49(2)

Clause 49(2) requires the contractor to have the works complete and free from defects as soon as practicable after the expiry of each defects correction period.

The clause contemplates that the engineer will make an inspection prior to the expiry of each such period and that he may also give instructions for repairs and defects to be made good during each period. It is not clear from the wording whether the obligation on the contractor is to complete all work within 14 days after the expiry of each period and to complete those works notified during the period as soon as practicable within the period. An alternative interpretation, which matches the requirement in the Fifth edition, is that the engineer must give his instructions within 14 days of the expiry of the defects correction period and the contractor's obligation is only to execute the work 'as soon as practicable'.

'Of whatever nature'

Clause 49(2) is not confined to repairs and defects for which the contractor is responsible. That is made clear by the phrase 'of whatever nature'. This is supported by the change in the Sixth edition to clause 51 empowering the engineer to issue variations during the defects correction period.

Remedy on contractor's failure

Clause 49(4) entitles the employer, in the event of the contractor's failure to do 'any such work as aforesaid', to carry out such work himself. The employer can recover the cost from the contractor if the work should have been carried out at the contractor's own expense.

It is not clear whether the phrase 'any such work as aforesaid' covers both outstanding works under clause 49(1) and remedial works under clause 49(2) or whether it applies only to the latter.

By tracing the phrase 'such work' through clause 46(3) there is a good case for saying in the Sixth edition that the remedy in clause 49(4) applies only to default under clause 49(2).

It has to be admitted that this particular logic does not apply to the Fifth edition although the provision there included the phrase 'required by the Engineer' and this would seem to have excluded outstanding work.

Action by employer

Whatever the scope of the employer's remedy it certainly is powerful in that it apparently operates before the expiry of the defects correction period. This limits the general proposition that the contractor has the benefit of the defects correction period to enter the site and put right his own defects. It would seem that if the contractor fails to undertake the required work 'as soon as practicable' the benefit can be extinguished.

However, before taking action with the intention of charging the contractor, the employer needs to consider carefully how he is going to justify his action if a formal dispute arises. By acting the employer will probably destroy the best evidence for his case and he will need to rely on records of the defects he has put right.

Temporary reinstatement

The lengthy clause 49(5) on temporary reinstatement in the Fifth edition is omitted from the Sixth. It covered the responsibility of the parties to highway authorities.

12.5 *Clause 50 – contractor to search*

Clause 50 obliges the contractor to carry out searches, tests and trials to determine the cause of any defect if so required in writing by the engineer. The clause is identical to that in the Fifth edition so the uncertainty remains as to whether its application is limited to the defects correction period or whether it has wider application.

There is nothing expressly in clause 50 to exclude its use during the construction period but the reference in its last line to clause 49 implies that it is intended for the defects correction period and the mention of clause 50 in clause 61 supports this.

It can also be said that clauses 36 and 38 are intended for use during the construction period.

Against this, however, is the point that each of the clauses 36, 38 and 50 serves a slightly different purpose. Clause 36 deals with tests which can be random or routine and ordered without sign of defect; clause 38 deals with uncovering work which has been put out of view; clause 50 deals only with searches in respect of patent defects.

Costs of searching

Clause 50 provides that the contractor is to bear the costs of searching etc. if the defects are found to be his liability; otherwise the employer bears the costs.

As with clauses 36 and 38 there can be problems in establishing how far the employer's liability extends when the searches and tests show some of his work to be acceptable but those searches and tests are necessitated by other defects for which the contractor is liable.

Works of repair

Where the contractor is liable, clause 50 obliges the contractor to repair and rectify at his own expense 'in accordance with clause 49'. This latter phrase presumably involves timescale.

Clause 50 says nothing on the ordering or execution of repairs where the contractor is not liable but the engineer has his powers under clauses 13, 51 and 62.

12.6 Clause 61 – defects correction certificate

Clause 61(1) provides for the issue by the engineer of a single 'Defects Correction Certificate' stating the date on which the contractor has completed his obligations to construct and complete the works to the engineer's satisfaction.

The clause makes it absolutely clear that where there is more than one defects correction period, as there will be of course with sections or with partial handovers, then it is only on the expiry of the last period that the defects correction certificate becomes due.

Purpose of the certificate

Clause 61(2) states that the issue of the defects correction certificate shall not be taken as relieving either of the parties of their liabilities to one another arising out of the contract.

This gives the certificate much lower status than similar certificates in other construction contracts where such certificates finalise certain matters so they cannot thereafter be disputed. All that the clause 61 certificate does is:

(a) establish the last date for the submission of the contractor's final account under clause 60(4), and
(b) terminate the defects correction period thus ending any remaining rights of the contractor to enter the site to remedy defects.

12.7 Clause 62 – urgent repairs

Clause 62 deals with accidents or similar events which may occur either during construction or during the defects correction period.

It provides that if the contractor is unable or unwilling to undertake such remedial work etc. as, in the opinion of the engineer or the engineer's representative, is urgently necessary then the employer may employ others to do such work.

The cost of the work so done is charged to the contractor if it is his liability; otherwise it is at the expense of the employer.

Chapter 13

Variations and claims

13.1 *Introduction*

This chapter covers clauses 51 and 52 which are given the sectional heading 'Alterations, Additions and Omissions' in the Conditions.

Clause 51 gives powers and procedures for ordering variations whilst clause 52, as in previous editions, deals with three separate subjects – the valuation of variations; the power to order dayworks; and notices for claims.

There are some important changes from the Fifth edition; not least that some of the difficulties of interpretation which have plagued these clauses through successive editions of the ICE Conditions may at last have been eliminated. Many users of the Conditions will be disappointed that more has not been done in this direction but that would have required radical departure from earlier editions and there is something to be said for the policy of the draftsmen in proceeding cautiously on issues in which dispute is inherent, however well drafted the provisions.

13.2 *Variations generally*

The courts will not ordinarily imply a term into a contract allowing one party the right to vary the content of the contract. Therefore if no express provision has been made for variations they can only be ordered by agreement and the terms are a matter for negotiation.

The near impossibility of avoiding variations in construction contracts has long been recognised and all standard forms have provisions which empower the contract administrator, be he architect or engineer, to order variations on behalf of the employer. That in itself, however, does not dispose of dispute and there are two broad issues which stand at the margins of every variation clause:

(a) Is a variation necessary for a particular item of work or change? Is it the contractor's responsibility or is it already allowed for in the contract rates or prices?
(b) Is a particular variation within the scope of the variation clause or is it something which should be valued outside the contract?

These can be illustrated by some recent civil engineering cases, all on the Fifth edition or variants thereof. Firstly, cases where the contractor successfully claimed a variation:

(a) *Yorkshire Water Authority* v. *Sir Alfred McAlpine & Son (Northern) Ltd* (1985) where it was held that the incorporation of a method statement into the contract made it a specified method of construction and the contractor was entitled to a variation order for an alternative method of working.
(b) *Holland Dredging (UK) Ltd* v. *Dredging & Construction Co. Ltd* (1987) where it was held that the contractor was entitled to a variation for measures in rectifying a shortfall in backfill to a sea outfall pipe arising from loss of dredged material at the dumping ground.
(c) *English Industrial Estates* v. *Kier Construction Ltd* (1991) where it was held that an engineer's letter instructing the contractor to crush all suitable hard arisings deprived him of the option of importing suitable fill and was therefore a variation.

For a case on the scope of the variation clause *Blue Circle Industries plc* v. *Holland Dredging Co. (UK) Ltd* (1987) is worth noting. In that case the contract was for dredging in Lough Larne but a separate quotation was given for the construction of an island to serve as a bird sanctuary. The island works were not a success and the employer sued for damages. The contractor applied for a stay in proceedings contending the island works were a variation and the arbitration clause of the contract applied. It was held the island works were outside the scope of the contract; the contractor would not have been obliged to undertake the work had he been unwilling; the construction of the island was not a variation but the subject of a separate agreement.

The *Blue Circle* case was somewhat unusual in that it was the contractor who was seeking to extend the scope of the variation clauses. Usually on a dispute over scope the contractor will seek to nullify a variation in order to be paid on a more favourable basis. This extract from the judgment of Lord Cairns in the old case of

Thorn v. *Mayor of London* (1876) illustrates the point:

> 'Either the additional and varied work which was thus occasioned is the kind of additional and varied work contemplated by the contract, or it is not. If it is the kind of additional or varied work contemplated by the contract, he must be paid for it, and will be paid for it, according to the prices regulated by the contract. If, on the other hand, it was additional or varied work, so peculiar, so unexpected, and so different from what any person reckoned or calculated upon, that it is not within the contract at all; then, it appears to me, one of two courses might have been open to him; he might have said: I entirely refuse to go on with the contract – *non haec in foedera veni*: I never intended to construct this work upon this new and unexpected footing. Or he might have said, I will go on with this, but this is not the kind of extra work contemplated by the contract, and if I do it, I must be paid a *quantum meruit* for it.'

13.3 Clause 51 – ordered variations

Clause 51(1) has two distinct limbs. The engineer 'shall' order any variation necessary for completion of the works and 'may' order any variation desirable for the completion or improved functioning of the works.

The clause gives a comprehensive list of items which may be included as variations. They can be categorised as follows:

- additions and omissions
- substitutions and alterations
- changes in quality, form, character, kind, position, dimension, level or line
- changes in any specified sequence, method or timing of construction required by the contract.

The clause further provides that variations may be ordered during the defects correction period.

Changes from the Fifth edition

The changes from the Fifth edition are:

(a) In clause 51(1)(a) the phrase 'is in his opinion necessary' replaces 'may in his opinion be necessary'. This suggests that the engineer is now required to have a greater degree of certainty as to what is and what is not 'necessary'.
(b) In clause 51(1)(b) 'may order any variation' has replaced 'shall have the power to order any variation'; 'the completion and/or improved functioning of the Works' has replaced 'satisfactory completion and functioning of the Works'. The first change is not thought to be significant; the second, by its reference to 'improved functioning' may give wider scope for variations.
(c) In the final sentence of clause 51(1), 'any specified sequence' has replaced 'the specified sequence' and the phrase 'required by the contract' has replaced the enigmatic bracketed phrase 'if any'. It is thought these changes are intended to emphasise that it is only changes to methods and timing etc. which are contractually binding which constitute variations. The argument, which never carried much weight, that the 'specified sequence' included the contractor's clause 14 programme has been put to rest.
(d) The final phrase of clause 51(1) 'and may be ordered during the Defects Correction Period' is new. This disposes of the much debated point on whether the Fifth edition provided for the issue of variations after the certificate of completion had been issued.

Necessary variations

The provision that the engineer 'shall', in the imperative, order any variation necessary for the completion of the works is of obvious benefit to the contractor but less so the the employer. It transfers the burden of dealing with impossibility from the contractor to the engineer, thereby eliminating, from these Conditions the proposition that the employer does not warrant that the works can be completed. It also puts the engineer under pressure to intervene when the contractor is in difficulties, for as the comment in chapter 6 on impossibility shows, there is a fine dividing line in construction between impossibility and difficulty.

There is also a lack of economic realism in the provision in that there is no point in the engineer being obliged to issue variations which are beyond the financial resources of the employer or which entail continuing with wasted expenditure on a project already rendered financially uneconomic.

Various clauses of the contract have a direct bearing on the engineer's duty to issue variations, for example:

clause 5 – documents mutually explanatory
clause 7 – further drawings and instructions
clause 12 – adverse physical conditions
clause 12 – engineer's instructions
clause 20 – excepted risks
clause 27 – street works
clause 62 – urgent repairs.

Other clauses can impose a duty in appropriate circumstances.

Desirable variations

The limitations on the power of the engineer to order 'desirable' variations may come from financial controls imposed by the employer – as now covered by clause 2(1)(b) and the corresponding appendix entry – or from lack of proximity of function as described above in the *Blue Circle* case.

Lack of proximity in the geographical sense may also be relevant if the varied work extends beyond the site boundaries although the new power to extend the site under clause 1(1)(v) could be helpful.

It is not thought that the words 'for any other reason' in clause 51(1)(b) are intended to confer a general power on the engineer to order variations. They probably mean no more than the words 'other than necessary' do in relation to clause 51(1)(a). It is doubtful, therefore, if the engineer can, for example, order omission variations purely as cost saving measures – although in reality it has to be said that such practice is commonplace.

The remedy for an aggrieved contractor, if satisfaction cannot be gained under the contract for unauthorised variations, is to sue the employer for breach.

Variation by agreement

Where additional work or cost is involved contractors do not always oppose variations but sometimes welcome them as an opportunity of increasing turnover and profitability. So a contractor will often accept a variation which is outside the scope of clause 51, as either necessary or desirable, and to the extent that he is prepared to do so, and the employer is of the same mind, such a variation serves as a substitute for a separate agreement. When this is done it is best that the variation is labelled a 'variation by agreement' to avoid any possibility of later dispute that it was unauthorised, unacceptable or void.

Range of variation items

The range of items listed in clause 51(1) as permissible variations is unusually wide compared to other construction contracts and the reference to changes in 'sequence method or timing' appears to give the engineer remarkable powers to interfere in, or control, the contractor's arrangements.

Taken to the extreme there is a tenable view that the engineer has power to order the contractor to accelerate or complete within a shorter time than that allowed in the contract – and indeed the new clause 46(3) on agreement of acceleration costs may be in recognition of this. But over-use, or abuse, of the range of items is not really a problem – except perhaps between the engineer and the employer – because the contractor will be entitled to payment under one clause or another.

The problem, if such it is, is the combination of the range of items and the engineer's duty under clause 51(1) to order 'necessary' variations. Thus in both the *Yorkshire Water* and *Holland Dredging* cases, the contractor was held to be entitled to payment because the specified method was no longer possible and the engineer was obliged to issue a variation to ensure completion.

Variations to be in writing

Clause 51(2) requires that all variations shall be in writing, with the provisions of clause 2(6) to apply to oral instructions. That clause is to the effect that if the contractor confirms an oral instruction of the engineer in writing and that confirmation is not contradicted in writing forthwith, the instruction stands as valid.

In practice, the process of issuing formal variation orders often falls well behind construction work either because there is disagreement on what constitutes a variation or because the engineer or employer, as a financial safeguard, insists that only fully priced variation orders should be issued. However, providing the contractor has a properly authorised written instruction, whatever it is called and whatever clause of the contract it is issued under, it will stand as a variation order if the appropriate circumstances apply.

Variation not to affect contract

Clause 51(3) provides:

(a) no variation shall vitiate or invalidate the contract, and
(b) the value of all variations is to be taken into account in ascertaining the contract price other than as necessitated by the contractor's default.

The first of these provisions seems to be no more than a precaution against legal challenge to the validity of the contract. Certain contracts outside the field of construction contain provisions to the opposite effect so that they are invalidated by any variation.

The second provision on entitlement to payment is changed from its counterpart in the Fifth edition with the addition of the phrase 'except to the extent that such variation is necessitated by the contractor's default'. A practical man might think these words unnecessary since it would seem to go without saying that the contractor would not be entitled to payment for a variation which was necessitated by his own default.

But contracts are taken to mean what they say and there would certainly have been a danger in leaving the clause unquantified. Thus, in the case of *Simplex Concrete Piles* v. *St Pancras Borough Council* (1958) an architect who assented to a contractor's proposals for overcoming defective work by his piling sub-contractor was held to have issued a variation which entitled the contractor to payment.

Acceptance of contractor's proposals

The *Simplex* case and others since have shown how careful engineers or architects must be in accepting contractor's proposals for either alternatives to the designed works or the rectification of defaults. Generally, if the engineer accepts an alternative to his design that becomes the specified method or requirement. If the method proves impossible or the requirement unsatisfactory the engineer is then bound to issue instructions under clause 13 or a variation under clause 51(1). The contractor is entitled to be paid because he is not contractually 'in default'.

In accepting proposals for the rectification of defects the engineer is not as exposed under the Sixth edition as he was under the Fifth but there is still need for caution to ensure that the wording of an approval does not inadvertently become an instruction which goes beyond the extent of the contractor's default.

Changes in quantities

Clause 51(4) is the same as clause 51(3) in the Fifth edition. It says that no written order is necessary for a change in quantities where such change is simply the result of the actual quantities being more or less than those stated in the bill of quantities.

Opinion is divided on whether the effect of this is to make every change in quantities a 'variation' although not, it should be noted, 'a variation ordered by the engineer'; or whether it merely confirms that changes in quantities are not to be regarded as variations. The matter is of some importance in connection with the phrase 'any variation' at the commencement of clause 52(2) as discussed below.

13.4 Valuation of variations – generally

There are three clauses in the Conditions dealing with the valuation of variations:

(a) clause 52(1) which covers the valuation of 'ordered' variations
(b) clause 52(2) which permits the re-valuation of rates rendered unreasonable or inapplicable by the nature or amount of 'any' variation
(c) clause 56(2) which permits the re-valuation of rates rendered unreasonable or inapplicable by changes in quantities.

Clause 52(1), it is thought, is confined to the valuation of the varied work itself whereas both clauses 52(2) and 56(2), in their references to 'any rates or prices', suggest they extend to items other than those directly in the variation or subject to the change of quantity.

There are strict notice requirements for activating clause 52(2) whereas both clauses 52(1) and 56(2) operate without notice and on the basis of consultation between the engineer and the contractor. This suggests that clause 52(2) is fundamentally different from either clause 52(1) or 56(2) and there are indeed some points to note:

(a) The responsibility for commencing the valuations in clauses 52(1) and 56(2) is put solely on the engineer. In clause 52(2), either the engineer or the contractor may commence action. This may appear a theoretical point since it is not unusual for the contractor to take the lead in all cases but it does emphasise that under 52(1) and 56(2), the contractor has rights and the engineer has duties.
(b) Both clauses 52(1) and 56(2) use bill rates as the starting points for revaluation and a fair valuation is only permitted in 52(1) when those rates cannot reasonably be used. There is no such restriction in clause 52(2) however and the engineer is empowered to fix rates he thinks to be reasonable and proper.

'Any variation'

Clearly there is more affinity between clauses 51(1) and 56(2) than either has with clause 52(2). But what clause 52(2) does is to act as a link between clauses 52(1) and 56(2) so that the principle of revaluation for change can apply to ordered variations as well as the ordinary quantity changes. What is not clear, as mentioned above, is whether such quantity changes are within the measure of 'any variation' and whether the contractor can use clause 52(2) as the clause for instigating a rate revaluation when the engineer is not disposed to act under clause 56(2).

The point is of some importance because whereas a rate revaluation under clause 56(2) can only take place when the 'increase or decrease of itself so warrants' there is no such restriction in clause 52(2). Thus a 'loss of profit' claim for reduced quantities might succeed under clause 52(2) but fail under clause 56(2).

Valuation of ordered variations

Clause 52 provides a complete code for the valuation of ordered variations and there is no room for the proposition sometimes put forward by contractors that the interaction, complexity or disruption of variations enables them to claim recovery of cost on what is called a global basis.

The well known cases put forward to support such claims, *Crosby* v. *Portland UDC* (1967) and *Merton* v. *Leach* (1985) both dealt with the interaction of different heads of claim and give no support to departure from the rules of clause 52 for variations. More recently in the case of *McAlpine Humberoak Ltd* v. *McDermott International Inc.* (1992) the Court of Appeal held on a lump sum contract that even a considerable number of variations did not alter the basis of the contract so as to render performance totally different from that envisaged originally and the contractual machinery could not be abandoned.

The remedy for the contractor in the the ICE Conditions, and in particular the Sixth edition, is to use to the full the provisions of clauses 52(1)(b) and 52(2).

Valuation of deemed variations

Various clauses permitting cost recovery, clauses 7 and 13 can be taken as examples, use wording of the kind 'any variation ... shall be deemed to have been given pursuant to clause 51'. The question is: does the valuation of these variations come under the rules of clause 52 or is recovery on a cost plus basis allowable?

The answer, it is suggested, is neither a straightforward yes or no. Where the clause specifically provides for the recovery of cost plus profit, as do clause 13 and others, it cannot be intended that rate comparability should apply since that might involve the contractor in further loss. However, where the clause is silent on the basis of recovery, as in clause 7(1), then valuation under clause 52 might well be appropriate.

13.5 Clause 52(1) – valuation of ordered variations

Clause 52(1) provides that the value of all variations ordered by the engineer in accordance with clause 51 shall be ascertained by the engineer after consultation with the contractor. If the engineer and contractor fail to agree, the engineer is required to fix any disputed rate or price and notify the contractor accordingly.

Clause 52(1) lays down three principles for ascertaining value:

(a) work of a similar character and executed under similar conditions to work in the bill of quantities is to be valued at bill rates
(b) work not of a similar character or not executed under similar conditions, or work executed during the defects correction period is, so far as reasonable, to be valued on the basis of the rates in the bill of quantities
(c) when neither of the above is applicable a fair valuation is to be made.

The logic in using bill rates and prices for valuations is that the contract itself is founded on these rates and since the contract contemplates variations it is fair to both parties that bill rates should be used in valuations. Some other construction contracts, it must be said, apply a different logic and require all variations to be valued on a cost plus basis.

One particular point to note in clause 52(1) is that reference is now made expressly to variations ordered in the defects correction period. This was always a contentious matter in the Fifth edition but it is now clear that bill rates no longer need apply.

13.6 *Clause 52(2) – engineer to fix rates*

Clause 52(2) provides for the re-valuation of bill rates and prices rendered unreasonable or inapplicable by the nature or amount of any variation. The re-valuation can apply to any rates or prices so affected and can be in either direction; that is an increase or a decrease.

The test to be used is expressed in the most general terms 'if the nature or amount of any variation relative to the nature or amount of the whole of the contract or any part thereof shall be such ... that any rate or price in the contract ... is by reason of such variation rendered unreasonable or inapplicable'. Some employers have attempted to regulate the use of the clause by issuing guidelines to their engineers suggesting that only variations above a certain percentage value, 10% or 15% or so, should trigger re-valuation of rates but that is to distort the purpose of the clause.

Giving notice

Clause 52(2) requires either the engineer to give the contractor, or the contractor to give the engineer, notice requiring a revaluation. Notice is to be given before the varied work is commenced, or as soon thereafter as is reasonable in all the circumstances.

The provision for notice from either the engineer or the contractor will normally operate such that the engineer looks for reductions and the contractor looks for increases but if a bundle of rates is altered it may well be that some go up and some go down.

The question of timing of notices under near identical provisions to clause 52(2) in the ICE Second edition was considered in the case of *Tersons Ltd* v. *Stevenage Development Corporation* (1963). In that case the phrase 'as soon thereafter as is practicable' was given wide interpretation although the decision of the Court of Appeal made clear that compliance with the notice condition was a condition precedent to payment. This question of timing was considered again the case of *Hersent Offshore* v. *Burmah Oil* (1978) under conditions similar to the ICE Fourth edition and there is was held that notice given after the varied work had been completed had not been given 'as soon thereafter as is practicable'.

Engineer to fix rates

There is no requirement in clause 52(2) for the engineer to consult with the contractor before fixing new rates but the practical need to do so is more apparent than in clauses 52(1) and 56(2) where he is required to consult. Without a breakdown of the contractor's rates it would be adventurous of the engineer to form an opinion of what was rendered unreasonable or inapplicable and a request for the breakdown and particulars should be the engineer's first step.

13.7 *Clause 52(3) – daywork*

Clause 52(3) in the Sixth edition does no more than give the engineer power to order 'additional or substituted' work to be executed on a daywork basis.

The detailed procedures for records etc. in clause 52(3) of the Fifth edition have been greatly simplified and transferred to clause 56(4) of the Sixth.

It should be noted that the engineer has no power to allow any of the original work to be executed on a daywork basis – that would amount to relieving the contractor of his obligation to construct the works at the rates and prices in his tender. And, as stated in clause 2(1) (c), the engineer has no authority to amend the terms of the contract or relieve the contractor of his obligations.

13.8 Clause 52(4) – notice of claims

Clause 52(4), although included with the variation provisions of the contract, details the requirements for the submission of all claims made under the contract. Such an important matter really deserves its own clause.

With the exception that the phrases 'or clause 56(2)' and 'in any event within 28 days' have been added into clause 52(4) (b), the whole of clause 52(4) is identical to its counterpart in the Fifth edition in both wording and arrangement.

Requirements of clause 52(4)

The provisions of clause 52(4) can be briefly summarised as follows:

(a) the contractor is to give notice of claims within a defined time
(b) the contractor is to keep such contemporary records as necessary
(c) the engineer may instruct the contractor to keep further records
(d) the contractor is to permit the engineer to inspect his records
(e) the contractor is to submit a first interim account
(f) the engineer may require the contractor to submit up-to-date accounts
(g) the contractor is to submit such accounts with accumulated totals
(h) the contractor is entitled to payment only to the extent the engineer is not prejudiced from investigating the claim
(i) the contractor is entitled to interim payment of amounts considered due by the engineer
(j) where the contractor has not substantiated the whole of the amount claimed he remains entitled to payment of any amount which is substantiated.

Scope of clause 52(4)

Clause 52(4)(a) covers claims which are effectively disputes on rates whilst clause 52(4)(b) covers claims for 'any additional payment pursuant to any clause of these Conditions'.

Pursuant means in accordance with, or in conformity with, so there is no doubt that all clauses which have a procedure for recovery of extra cost are covered by clause 52(4)(b). Prominent amongst such clauses are clauses 7, 12, 13, 31 and 40. What is less clear is whether clauses which permit recovery of extra cost or expense but have no procedure or reference to clause 52(4) are also covered. For example, is the contractor bound by clause 52(4) if he:

(a) makes a claim under clause 17 for incorrect setting-out data; or under clause 32 for fossils?
(b) makes a claim under clause 20 for an excepted risk; or under clause 29 for unavoidable nuisance?
(c) makes a claim for tests under clause 36; or searches under clause 38 or clause 50?

In some of the above cases there may be instructions under clause 13 or variations under clause 51 which will bind the contractor but otherwise it is doubtful if he is bound.

Extra-contractual claims

Where the contractor is almost certainly not bound by clause 52(4)(b) is in respect of claims for breach for which no provision is made in the contract – sometimes called extra-contractual claims. For example, a claim for delay caused by the engineer's failure to 'attend' under clause 38; or a claim for acceleration costs caused by the engineer's failure to operate clause 44. Although such claims have to go through the clause 66 procedure if they become the subject of disputes between the contractor and the employer the wording of that clause 'in connection with or arising out of the contract' gives much greater scope than clause 52(4).

Alternative claims for damages

The question of the scope of clause 52(4) is of particular interest insofar that its requirements on notices etc. have a limiting effect on the contractor's rights of claim. This raises the further question – can the contractor ignore the requirements of clause 52(4) for contractual claims and submit alternative claims for damages?

The answer is most certainly that he cannot unless the claims genuinely are damages for breach; and many contractual claims are not for breach – clause 12 for unforeseen conditions being the most obvious example. Even when contractual claims are for breach, such as late information under clause 7 or late possession under clause 42, there are two schools of thought on whether observance with clause 52(4) is essential or whether the contractor has alternative rights of claim.

Those who argue that the contractor has alternative rights and it matters not that the breach is covered contractually do so on the proposition that common law rights can only be excluded by express terms and no such terms are to be found in the ICE Conditions. They gain some support from the words of Mr Justice Vinelott in *Merton* v. *Leach* (1985), a case on a JCT 63 building contract. He said:

> 'In either case the contractor can call on the architect to ascertain the direct loss or expense suffered and to add the loss or expense when ascertained to the contract sum. The contractor will then receive reimbursement promptly and without the expense and delay of a claim for damages. But the contractor is not bound to make an application under clause 24(1). He may prefer to wait until completion of the work and join the claim for damages for breach of the obligation to provide instructions, drawings and the like in good time with other claims for damages for breach of obligations under the contract. Alternatively he can, as I see it, make a claim under clause 24(1) in order to obtain prompt reimbursement and later claim damages for breach of contract, bringing the amount awarded under clause 24(1) into account.'

But it may be significant that JCT 63 did have a clause, as does JCT 80, saying:

> 'The provisions of this Condition are without prejudice to any other rights and remedies which the contractor may possess,'

Those who argue against the contractor having alternative rights do so on the basis that the contract is an agreement, freely entered into, and procedures which have been expressly so agreed cannot unilaterally be discarded or disregarded.

Until there is a contractor brave enough to put this matter to the test of legal decision it is obviously prudent for contractors to comply with clause 52(4) unless circumstances are such that submission of a late claim without notice cannot be avoided.

Meaning of notice

Under clause 52(4) the contractor is required to give 'notice' in writing of his intention to claim additional payment. The meaning of 'notice' as used in clause 52 was one of the questions considered in the *Tersons* v. *Stevenage* case. The Court of Appeal upheld the decision of Mr Justice Roskill who had said:

> 'I think it is sufficient that the notice should specify that a claim is being made, provided that the notice identifies in general terms the nature of the additional work to which the claim will relate when it is ultimately precisely formulated.'

13.9 Clause 52(4) – detailed provisions

Rates and prices

Clause 52(4) (a) applies when the contractor is dissatisfied with a rate or price set by the engineer under clause 52(1) (variations) or clause 56(2) (changes in quantities).

The contractor is required to give notice within 28 days after notification of the rate.

Additional payments

Clause 52(4) (b) applies to claims for additional payments other than for rates and prices.

The contractor is to give notice of his intention to claim 'as soon as may be reasonable and in any event within 28 days after the happening of the events giving rise to the claim'. Upon the happening, the contractor is to keep contemporary records to support his claim.

The 28 day requirement in this clause is new to the Sixth edition and contractors need to be alert to this.

Contemporary records

Clause 52(4)(c) entitles the engineer, on receipt of a claim notice, to instruct the contractor to keep contemporary records. It requires the contractor to keep such records; to make them available to the engineer for inspection; and to supply copies as instructed.

Submission of accounts

Clause 52(4)(d) requires the contractor to send to the engineer a first interim account as soon as reasonable after giving notice. The account is to give full and detailed particulars of the amount claimed and the grounds of claim.

Thereafter at such intervals 'as the engineer may reasonably require' the contractor shall send further up-to-date accounts with accumulated totals and any further grounds of claim.

It is disappointing that the Sixth edition has left the initiative for requiring up-to-date accounts with the engineer. This could have been placed on the contractor to link in with clause 60(1)(d) on interim applications for payment. As it is, engineers are often reluctant to do anything which appears as encouragement to contractors to submit bigger and better claims.

Failure to comply

Clause 52(4)(e) is something of a saving provision for the contractor for it preserves the contractor's entitlement on any claim, which has not been submitted in accordance with clause 52(4), to the extent that the engineer has not been prevented or prejudiced from its investigation.

Contractors should beware of taking this as general freedom to submit post-completion or other late claims, particularly of the global type. It was made quite clear in both *Crosby* v. *Portland* (1967) and *Merton* v. *Leach* (1985) that such claims can only succeed when the provisions of the contract have been complied with but interaction of events prevents assessment under individual heads of claim.

Interim payments

Clause 52(4)(f) entitles the contractor to interim payment of amounts the engineer considers due on claims.

The clause further provides that, if the particulars submitted by the contractor are insufficient to substantiate the whole of the claim, the contractor shall be entitled to payment for as much as the particulars may substantiate.

What this is saying is that, the engineer cannot refuse to certify anything on a claim because the documentation falls short of proving the full amount claimed. The engineer remains under a duty to certify as much as has been substantiated.

Global claims

A further word of warning here on global claims. These suffer from the defect that if any part of the amount claimed is not the responsibility of the employer, the whole of the amount claimed is potentially invalidated. This is what Mr Justice Vinelott said in the *Merton* case:

> 'I find it impossible to see how the calculation can be treated as even an approximation for a claim, whether or not rolled up ... the calculation in effect relieves Leach from any burden of additional costs resulting from delays in respect of which Leach is not entitled to any extension of the completion date.
>
> The calculation might be appropriate to a claim for a *quantum meruit* or for a partial *quantum meruit* to reflect additional expense which, by reason of default on the part of Merton, cannot be calculated in accordance with the machinery provided. But as I have said a claim for a partial *quantum meruit* is not one on which in my judgment Leach is entitled to pursue.'

Chapter 14

Property and equipment

14.1 Introduction

This chapter examines clauses 53 and 54, vesting of contractor's equipment and vesting of goods and materials not on site.

Clause 53 in the Sixth edition has been reduced to about one tenth the length of clause 53 in the Fifth and it no longer purports to establish as wide a range of employer's rights.

Clause 54 is virtually unchanged.

14.2 Purpose of clause 53

Most construction contracts have clauses which seek to avoid the difficulty for the employer that if the contractor's business fails his assets fall into the control of a trustee, receiver or the like and will probably be disposed of by prompt sale. The purpose of clause 53 in the ICE Conditions and similar clauses elsewhere is to establish the contractor's equipment, goods and materials as the property of the employer as soon as they are brought on site. This is to protect them against claims from the contractor's legal successors or sub-contractors so that they remain available for use in completing the works.

The Fifth edition in its very complex provisions attempted to give the employer comprehensive cover by including any plant – which by definition meant construction plant, temporary works, materials etc. – which was either:

(a) owned by the contractor
(b) owned by a company in which the contractor had a controlling interest
(c) hired by the contractor
(d) brought on site by sub-contractors.

The provisions were vulnerable, however, to the doctrine of privity of contract and conditions of plant hire and were only truly effective when the contractor himself had full ownership.

A well publicised building case under JCT Conditions which illustrated the point was *Dawber Williamson* v. *Humberside County Council* (1979).

A roofing sub-contractor brought slates on to site for which the contractor was paid by the employer. Before the slates were fixed the contractor went into liquidation. The sub-contractor attempted to retrieve the slates but the employer prevented this and claimed ownership. The sub-contractor then successfully sued the employer for their value. In considering the clause in the main contract which said that when the value of any goods has been included in any certificate under which the contractor has received payment, the goods shall be the employer's property, Mr Justice Mais said:

> 'In my judgment, this presupposes there is privity between the defendants and the sub-contractor, which there is not in the present instance, or the main contractor has good title to the material and goods. If the title has passed to the main contractor from the sub-contractor, then this clause has force.'

Retention of title clauses

It should perhaps be added here that the law treats differently goods acquired under a sale of goods from goods acquired under a contract for work and materials (supply and fix). Goods acquired under a sale come within the scope of the Sale of Goods Act 1979. Under section 17 of that Act property is transferred at such time as the parties intend it to be transferred and regard is to be made, in ascertaining such intention, to the terms of the contract, the conduct of the parties and the circumstances of the case.

To avoid title passing before payment is made, many contracts for sale of goods often have clauses granting possession and even use to the buyer but retaining ownership with the seller until payment. These are sometimes called Romalpa clauses after the case of *Aluminium Industries* v. *Romalpa Aluminium Ltd* (1976).

See also the comment on ownership of goods on site and off site in section 2 of chapter 18.

14.3 Clause 53 in the Sixth edition

Clause 53 in the Sixth edition deals only with equipment, materials etc. owned by the contractor.

Under the definitions in the Sixth edition, equipment covers what was previously called 'plant' in the sense of constructional plant.

Clause 53(1) now provides that all equipment, temporary works, and other goods or materials 'owned by the contractor' shall:

(a) when on site be deemed to be the property of the employer, and
(b) not be removed without the written consent of the engineer.

There is no longer any attempt to control hired plant or the equipment of sub-contractors and many of the difficulties faced by engineers with the old clause 53 have thankfully been removed.

Engineer's consent to removal

Clause 53(1) further provides that the engineer's consent to removal shall not be unreasonably withheld where the items involved are no longer immediately required for completion of the works.

This permits the contractor a fair deal of flexibility in moving equipment on and off the site and perhaps now that the clause generally has been reduced to working proportions a greater attempt will be made to operate it in practice.

Liability for loss or damage

Clause 53(2) provides that the employer shall not be liable for any loss or damage to the contractor's equipment, goods or materials except as mentioned in clause 22 (damage to persons and property) and clause 65 (war clause).

The purpose of this is to distinguish between ownership and risk. Although the various items have been deemed to be the property of the employer under clause 53(1) it is not intended that the employer should be liable for their loss or damage. They remain the full responsibility of the contractor.

The reference to clause 22 applies only to the exceptions in that clause. Under the Fifth edition the exception in clause 53(a) was to clause 20 (care of the works) not clause 22.

It is difficult to avoid the conclusion that both the exceptions under clause 22 and the excepted risks under clause 20 should be excluded from clause 53(2). The intention and implications of the change are not obvious unless it is simply that clause 22 has an express indemnity by the employer, and clause 20 does not.

In any event the exclusions may not be as effective as they appear at first sight. The comments in chapter 6 under 'the effects of clause 12', following the *Humber Oils* case show the employer can have unexpected liabilities for the contractor's equipment.

Disposal of contractor's equipment

Clause 53(3) allows the employer to sell, or dispose of, any of the contractor's equipment, goods or materials not cleared from the site within a reasonable time after completion of the works.

The reference to clause 33 in clause 53(3) is made because clause 33 expressly states the contractor's obligation to clear away on completion.

The employer is entitled under clause 53(3) to offset the costs of disposal against any proceeds before paying the balance to the contractor. If the balance is negative the employer could sue for breach of clause 33 or exercise set-off against a certificate. It is unlikely, on the wording, that a negative balance could be regarded as a sum 'to which the employer is entitled under the contract' for the purposes of the engineer's final certificate under clause 60(4).

14.4 *Vesting of goods and materials not on site*

Clause 54 provides a mechanism for the contractor to be paid for goods and materials not on site where:

(a) they are listed in the appendix to the form of tender
(b) they have been manufactured and are ready for incorporation into the works
(c) they are the property of the contractor; or the contract of supply makes provision for them to pass unconditionally to the contractor.

'If the Engineer so directs'

Clause 54(1) is the same as in the Fifth edition except there is an added provision that the contractor 'shall if the Engineer so directs' transfer to the employer the property in goods and materials to which the clause applies. This may be a useful provision if it can be seen that the contractor is in financial difficulties and likely to fail. It is not clear if it has any wider purpose.

Certification and payment

Operation of the provisions of clause 54 allows the engineer to certify the amounts due for off-site goods and materials under clause 60(1) (c) in interim certificates and obliges the employer to pay under clause 60(2) (b).

Action by contractor

As evidence of ownership of off-site goods and materials, the contractor is required by clause 54(2) to take, or cause his supplier to take, action in:

(a) providing the engineer with documentary evidence that property has vested in the contractor
(b) marking and identifying goods and materials to show:

  (i) their destination is the site
  (ii) they are the property of the employer
  (iii) to whose order they are held

(c) setting aside and storing to the satisfaction of the engineer
(d) sending the engineer a list and schedule of values
(e) inviting the engineer to inspect.

Engineers should ensure that these requirements are strictly fulfilled because of the inevitability of conflicting claims to ownership if there is a business failure somewhere along the lines of supply or the contract chain.

Vesting in employer

Clause 54(3) recognises that the employer obtains full ownership under clause 54 – not simply deemed ownership as under clause 53.

It, therefore, makes necessary provision for:

(a) rejection of goods and materials not in accordance with the contract
(b) ownership to revest in the contractor upon rejection
(c) the contractor to be responsible for loss or damage
(d) the contractor to take out additional insurance.

Lien on goods or materials

Clause 54(4) is intended to prevent the contractor, sub-contractor or any other person holding the goods or materials which have vested in the employer against sums they allege to be due.

This may be legally effective against the contractor or his legal successors but it is of limited practical value against third parties.

Delivery of vested goods and materials

The purpose of clause 54(5) is to give the employer the right to obtain delivery of, or take possession of, goods or materials after determination of the contractor's employment under clause 63 or 'otherwise'.

'Otherwise' may refer to common law determination or termination of the contractor's employment by agreement. Whatever the intention, there is here a good reminder that novation agreements need to be carefully drafted in respect of off-site goods and materials since the new contractor will not automatically be bound by or have the benefit of the old contractor's contracts of supply.

Incorporation into sub-contracats

The purpose of clause 54(6) is probably to ensure that the contractor's rights against sub-contractors are no less than the employer's against the contractor so that in the event of default in the contractual chain the employer's title is not lost.

But such clauses for 'equivalent provisions' are open to the criticism that only rarely will the contractor deal with sub-contractors on the same basis as he deals with the employer and there is usually no corresponding or similar contractual structure to graft the 'equivalent provisions' into.

Chapter 15

Measurement

15.1 *Introduction*

This chapter covers clauses 55, 56 and 57 dealing with measurement.

The three clauses remain much the same as in the Fifth edition but the procedure for daywork is moved from clause 52 into clause 56 and the new wording of clause 57 on exceptions to the Standard Method of Measurement is wider than in the Fifth edition.

Remeasurement contract

It is beyond any doubt that the ICE Conditions, both Fifth and Sixth editions, are remeasurement contracts and not of the lump sum type. This is evidenced by the lack of any link between the tender total and the contract price as defined in clause 1. The process of measurement therefore is not undertaken to produce an adjustment to the tender total; it is to arrive at the contract price by ascertainment of the actual quantities for which the contractor is entitled to be paid.

Legal precedents

There is very little by way of legal precedent to assist engineers and other users of the ICE Conditions in the correct interpretation of the measurement clauses of the contract. In the last half century only a few cases of any note have been reported:

Bryant & Son v. *Birmingham Hospital Saturday Fund* (1938)
Dudley Corporation v. *Parsons & Morrin Ltd* (1957)
A. E. Farr v. *Ministry of Transport* (1965)

and the decisions in these cases rely so heavily on the particular facts that it would be unwise to draw any general conclusions.

That is not to say that disputes on measurement are uncommon – far from it. If figures were available, which they are not, they might reveal that measurement disagreements form the majority of disputes referred to arbitration. But usually such disputes will be resolved by the arbitrator's findings of facts and whilst appeals to the courts may be allowed on points of law, they are not allowed on findings of facts.

Standard methods of measurement

It follows from the above that there is very little about measurement which can be said to be certain. And from that it follows that the use of a standard method of measurement which is clear in its rules is absolutely essential to any remeasurement contract.

Although the ICE Conditions refer specifically to the 'Civil Engineering Standard of Measurement' there is no reason why other methods should not be used with the conditions and the appendix to the form of tender allows for this. The problems come, more often than not, with well intentioned attempts at *ad hoc* modifications to standard methods which go wrong. And usually in such circumstances no amount of scrutiny of the conditions of contract will be of any assistance in providing specific answers to queries. The best the conditions can do is to provide a framework for the engineer to make decisions and disputes to be resolved.

15.2 Clause 55 – quantities

Clause 55 in the Sixth edition is identical to that in the Fifth.

Estimated quantities

Clause 55(1) states that the quantities in the bill of quantities are estimated and are not to be taken as the actual and correct quantities. This is, in effect, a statement by the employer that he does not warrant the quantities to be correct. There is, therefore, no place for an implied term that the quantities are correct.

It would require exceptional circumstances for the contractor to have any case against the employer for breach of contract for incorrect quantities. It would need to be shown that the quantities were not truly 'estimated' but were either recklessly or deliberately inaccurate such that there was negligence in the first case or misrepresentation in the second.

Risk on quantities

Although the employer does not warrant the quantities to be correct, the principle of remeasurement and the various provisions of the contract for rate adjustment when the quantities are not as billed places most of the risk on quantities firmly with the employer. But the contract does not go so far as to place all the risk with the employer and the contractor carries the risk that quantity changes may affect his return on the contract.

For example, contractors sometimes complain of loss of profit on items which have disposal value where the actual quantities fall short of those billed but the provisions for rate adjustment do not apply. Removal of topsoil is a typical item. The contractor's argument runs along these lines – the bill rate is for the work itself; the expected resale profit has been put to reducing preliminaries; the employer is receiving the benefit of the expected profit by such reduced preliminaries and should pay for it.

It is difficult to find any certain contractual route out of this for the contractor. The problem has arisen through innocent rate manipulation – the true bill rate for topsoil removal should have been negative. The employer is not at fault; the contractor has taken a risk which has back-fired; the loss lies where it falls. The best the contractor can do is to argue his case under clauses 52(2) or 56(2) for rate revaluations.

Correction of errors

Clause 55(2) deals with errors or omissions in the bill of quantities. It states that any error or omission shall not vitiate the contract nor release the contractor from his obligations and such error or omission shall be corrected and the work actually carried out shall be valued in accordance with clause 52.

It further states that there shall be no rectification of any errors, omissions or wrong estimates in the descriptions, rates and prices inserted by the contractor in the bills.

Clause 55(2) has to be read in conjunction with clause 57 (method of measurement) for the test for errors or omissions will be conformity with the rules of the stated method of measurement.

The clause is not as clear as it might be but its purpose is fairly obvious; if any item or element of work is wrongly billed or omitted – say, perhaps, the omission of a whole bridge abutment – that work is to be constructed, measured and valued.

Without this clause there might be some faint argument that the contractor was not entitled to have the work measured and valued. It might be said that under clause 5 the engineer is to explain discrepancies on the basis that the contract documents are mutually explanatory of one another and under clause 11 the contractor's rates and prices are to cover all his obligations under the contract. Clause 55(2), however, dispels such contentions and the drawings and specification are given precedence over the bill of quantities in determining the contractor's obligations and what shall be measured.

Valuation in accordance with clause 52

Clause 55(2) creates a few problems of interpretation in its reference to errors or omissions being 'corrected' by the engineer and the value of work 'actually carried out' being ascertained in accordance with clause 52. This appears to place the status of variations on the correction of such errors and omissions and adds weight to the argument that clause 51(4) does intend changes in quantities to be regarded as variations.

However it is contrary to clause 55(2) to regard as a variation something already deemed by the clause to be the contractor's obligation – as the construction of the bridge abutment in the example above. The correction of an error in the bill of quantities could never properly be described as a variation, nor would it give the contractor a right to extension of time.

Presumably, the reference to clause 52 should be taken principally as means to invoke the rules of clause 52(1) in using bill rates as the basis for valuation. However it is possible to take the purpose of the reference to clause 52 if strict application of the words 'actually carried out' is used as invoking clause 52(2) (effects of variations on existing rates).

No rectification of errors by contractor

The proviso in clause 52(2) that there is to be no rectification of errors made by the contractor emphasises that the purpose of the clause is to deal only with errors and omissions in the bills as issued – not as returned when priced.

The contractor's rates and prices are, therefore, naturally excluded. The 'descriptions' referred to are preliminaries and the like where the method of measurement allows the contractor flexibility to enter his own items. The contractor cannot obtain, after acceptance of his tender, any correction of these.

If there is an error in computation between the rates and grossed-up totals there is little doubt that under the ICE Conditions the rates stand as effective notwithstanding the fact that the employer may be misled into acceptance of the tender. Some commentators have suggested that to maintain the equity of the situation the rates should be adjusted backwards from the totals but there is nothing in the contract to support this, and if the employer has suffered a loss his remedy is against the engineer for negligence in checking the bills.

15.3 Clause 56(1) – measurement and valuation

Clause 56(1) is a straightforward statement that the engineer shall determine the value of the work done by 'admeasurement'. The clause provides the contractual link between the measurement provisions of clause 55 and the payment provisions of clause 60.

'Except as otherwise stated'

These words are included in clause 56(1) to cover dayworks, expenditure under prime cost sums and other items which by the method of measurement are not capable of being measured.

'Admeasurement'

This is an old fashioned word meaning the process of applying a measure to ascertain dimensions.

The burden is clearly on the engineer to undertake the admeasurement but that does not absolve the contractor from a similar responsibility. Under clause 60(1)(a) the contractor is to submit monthly estimates of contract value and under clause 60(4) is to submit a final account with supporting documentation showing in detail the amounts he considers due.

The scheme is that the engineer measures on behalf of the employer, and the contractor measures for himself. If the contractor neglects to measure he is not expressly in breach of contract or disqualified from payment – he would jeopardise his right to interim payments and he would be reliant on the engineer's figures for the final valuation. However, if the engineer neglects to measure that is failure to carry out his duties under the contract. That might amount to a technical breach of contract although it is unlikely that the contractor could prove his loss; more likely it would amount to negligence as between the engineer and the employer.

Work done 'in accordance with the contract'

The engineer is only required to measure and value work done 'in accordance with the contract'.

Taken together with clause 39 (removal of improper work and materials) it confirms that under the ICE Conditions there is no power for the engineer to accept sub-standard work on a reduced value basis. Some standard forms have brought in express terms to deal with this but they can perhaps be criticised for formalising bad workmanship. In practice, under the ICE Conditions, engineers and contractors sometimes agree to artificial variations in order to downgrade both specification and prices but a formal arrangement outside the contract between the employer and the contractor is more satisfactory.

15.4 Clause 56(2) – increase or decrease of rate

Clause 56(2) remains identical to clause 56(2) of the Fifth edition which produced such a furore when introduced in 1973.

It provides that, should the actual quantities be greater or less than those in the bill of quantities and if, in the opinion of the engineer, such increase or decrease of itself shall so warrant, the engineer shall, after consultation with the contractor, determine an appropriate increase or decrease of any rates or prices rendered unreasonable or inapplicable.

In short it can be expressed thus – a change in quantities may give cause for revision of the contractor's bill rates. And this, of course, was the reason for concern for it appeared to undermine the very basis of the ICE Conditions. It was feared that, given express provisions for rate revisions, contractors would seek to maximise every opportunity to claim higher rates and engineers would use every instance of increased quantities to fix lower rates. But in fact the clause is not an open cheque for change and it was soon realised that the key phrase 'of itself' restricts its operation to rational adjustments wholly compatible with the balance of risk in the contract.

'Increase or decrease of itself'

The precise meaning of 'increase or decrease of itself' is elusive but it is generally understood to apply to a change of quantity which itself changes the method of working or the economics of a working method. It is not thought to apply to a quantity change which does no more than alter the overall profitability of the contract by yielding more or less volume of a work item.

'After consultation with the contractor'

These words repeat the phrase used in clause 52 on the valuation of variations. In both clauses the engineer is drawn into rate fixing which requires knowledge of the contractor's pricing assumptions to be effective. Consultation with the contractor is, therefore, imperative.

'An appropriate increase or decrease'

It is not always true that an increase of quantity will reduce the costs of production and signify a rate reduction. Consider the costs of disposal of off-site material which can increase dramatically once the contractor's planned tipping facility has been exhausted. In every case it is necessary to examine the economics of working.

It is suggested that what the clause contemplates is the discovery of the change of rate necessary to restore the same level of unit profitability to an item as in the tender rate.

There is no justification whatsoever for increases or decreases in quantities to be the excuse for building up new rates from scratch. The starting point must always be the bill rates.

'Any rates or prices'

The generality of this phrase suggests that it is not only the rates and prices of the items which have specifically increased or decreased which may be reviewed. Suppose for example the contractor intended to use lorries carting topsoil off the site for return loads of imported fill, with the cost of transport spread over both rates. Then if the volume of topsoil decreased that would directly increase the unit cost of imported fill.

'Rendered unreasonable or inapplicable'

It is easy enough to see what is meant by rates rendered 'inapplicable' – the assumptions made in pricing at tender stage no longer apply. But rates rendered 'unreasonable' is a vaguer concept.

Contractors say that to be asked to do more of a loss making item is unreasonable; but in truth it is not the rate which is rendered unreasonable but the amount of the loss and correcting that is outside the scope of clause 56.

More difficult to refute is the proposition put forward by engineers that, where an increase in quantity provides the contractor with an opportunity to change his method of working and reduce his unit rates, it is reasonable that there should be a reduction in the rates, whether or not the contractor actually changes his methods. In theory the view may be correct but in practice the emergence of increased quantities is often a gradual process. It certainly is not reasonable for the engineer to use hindsight if neither he nor the contractor had the benefit of foresight.

15.5 Clause 56(3) – attending for measurement

The provisions of clause 56(3) have to be read in conjunction with clause 38(1) – examination of work before covering-up.

Clause 38(1) requires the contractor to give notice to the engineer before covering up permanent work and the engineer is required to attend for the purposes of examining and measuring.

Clause 56(3) requires the engineer to give notice to the contractor when he intends to measure any part of the works.

Contractor's failure to attend

If the contractor fails to attend for measurement clause 56(3) provides that any measurements made by the engineer are to be taken as the correct measurement of the work. This might seem to bind an arbitrator, although if there was clear evidence that the engineer had made a gross mistake, it could probably be corrected.

15.6 Daywork

Daywork under the Fifth edition was covered by clause 52(3) which sets out both the engineer's power to order daywork and the procedure to be used.

In the Sixth edition the power to order daywork remains in clause 52(3) but the procedures and requirements on records are in clause 56(4).

The requirements for records are much simplified and gone are the old references to exact lists in duplicate delivered each day. Instead, clause 56(4) requires the contractor to furnish to the engineer such records, receipts etc. as may be necessary for proof of cost.

Engineer to direct

Engineers should note particularly that the clause provides that such returns shall be in the form and delivered at the times 'the Engineer shall direct'. This is quite a change from the Fifth edition where the contractor was under a duty to make returns as specified in clause 52(3) and in the event of failure was entitled only to such payment as the engineer considered fair and reasonable.

Now it is the engineer who must take the lead in directing the procedure and if the engineer fails to direct he will not be able to complain about the contractor's behaviour.

Agreed within a reasonable time

The provision that daywork returns shall be agreed within a reasonable time places a duty on both the contractor and the engineer. If the contractor fails he is the loser, but if the engineer fails the employer is the loser.

The contractor is entitled to include daywork values in his monthly statements and failure by the engineer to certify would leave the employer liable for interest charges.

15.7 Clause 57 – method of measurement

Clause 57 follows closely the wording of the Fifth edition in deeming the bill of quantities to have been prepared and measurements made in accordance with the Standard Method of Measurement. The opening words are different, however, and seem to provide greater scope for exceptions.

In the Fifth edition unless statements or descriptions in the bill of quantities showed to the contrary the bills were deemed to be prepared to the Standard Method. In the Sixth edition that exception remains but two more are added:

> 'unless otherwise provided in the Contract', and
> 'or any other statement to the contrary'.

Contractors will have to beware of this and estimators will need to examine all the contract documents.

Chapter 16

Provisional sums and prime cost items

16.1 *Introduction*

This chapter covers clause 58 which deals with the use of provisional sums and prime cost items.

In the Fifth edition, clause 58 contained within its text definitions of provisional sums and prime cost items but in the Sixth edition the definitions are given in clause 1(1).

16.2 *Provisional sums*

Characteristics

By the definition in clause 1(1)(l) a provisional sum has three characteristics:

(a) it is included and designated as a provisional sum in the contract
(b) it is a specific contingency for the execution of work or the supply of goods
(c) it may be used in whole or in part or not at all at the discretion and direction of the engineer.

On this there is no change from the Fifth edition.

Use of provisional sums

Clause 58(1) states that for every provisional sum the engineer may order the work to be done, or goods to be supplied, by either the contractor or a nominated sub-contractor (or both).

Where work is done by the contractor it is valued in accordance with clause 52; where work is done by a nominated sub-contractor it is valued in accordance with clause 59.

The Fifth edition had identical provisions in clause 58(7).

Non-use of provisional sums

Neither the Fifth nor the Sixth editions entitle the contractor to an instruction or variation on the non-use of a provisional sum and the contractor has no claim in respect of non-use.

Provisional items and quantities

Bills of quantities not infrequently contain 'provisional items' and 'provisional quantities' for earthworks and other matters of uncertainty. These do not come within the scope of clause 58.

The contractor is entitled and obliged to carry out work in provisional items and quantities if such work arises. The engineer cannot order the work to be carried out by a nominated sub-contractor.

Programming of provisional sums

The vexed question of whether a contractor is obliged to include in his programme work in a provisional sum has not been satisfactorily resolved. Nor has the follow-on question – is the contractor entitled to an extension of time for carrying out work in a provisional sum?

It is sometimes argued that a contractor should not be expected to programme work which may never be carried out and the ordering of any work under a provisional sum should be regarded as extra.

It is suggested that this argument is flawed. The work in a provisional sum is part of the contract work and it is not a variation when it is ordered. Moreover there are no provisions in clause 44 for extending time for the work in provisional sums. Probably the best the contractor can do is to argue that an instruction to carry out work under a provisional sum is also an instruction under clause 13. Then to the extent that the timing of the instruction or its contents delays or disrupts his arrangements he is entitled to recover any extra cost which could not have been foreseen at the time of tender. He is also entitled to an extension of time.

The contractor is, of course, on far stronger ground if the value of work exceeds the provisional sum. It can logically be argued that the provisional sum is the limit of what the contractor has undertaken to perform and any excess must be in the nature of a variation.

16.3 *Prime cost items*

Characteristics

The definition in clause 1(1)(k) gives a prime cost item two firm characteristics:

(a) it is an item in the contract containing wholly or in part an item referred to as prime cost, and
(b) it will be used for the execution of work or the supply of goods or services.

Use of prime cost items

Clause 58(2) states that in respect of a prime cost item the engineer may order either or both of the following:

(a) the contractor to employ a nominated sub-contractor in accordance with clause 59, or
(b) the contractor himself to carry out the work or provide the goods and services himself – but only if the contractor consents so to do.

This is the same as in the Fifth edition.

Note that by its definition a prime cost item 'will be used'. The engineer must therefore attempt in the first instance to follow either the nominated sub-contractor route or the contractor by consent route. Only if the contractor declines to take on the work himself or objects under the provisions of clause 59 to nomination can the engineer omit the work by variation under clause 51.

The use in clause 58(2)(b) of the phrase 'the contractor himself' is not thought to exclude sub-contracting under clause 4 as the means of execution of the work.

Payment for prime cost items

Where a nominated sub-contractor is used, payment is in accordance with the provisions of clause 59(5) under which the contractor gets:

(a) the net amount due to the nominated sub-contractor under the contract, plus
(b) any item in the bill of quantities for labours, plus
(c) the percentage for other charges and profit in the bill of quantities or in its absence the percentage in the appendix.

Where the contractor undertakes the work or provides the goods and services himself he is entitled to be paid the price he has quoted. In the absence of such a quotation the value is determined in accordance with clause 52.

Where the engineer, by variation order, omits a prime cost item, the contractor remains entitled to his percentage for charges and profit.

The value of a prime cost item can exceed the amount in the bill of quantities without any need for a variation but the contractor may be entitled to an extension of time under clause 44(1) (b).

Programming of prime cost items

The contractor should include in his programme work to the full value of any prime cost item. But, of course, the contractor may not know when he submits his clause 14 programme the time requirements of nominated sub-contractors who may still have to be selected. If there is a problem accommodating the time requirements of a nominated sub-contractor into the contract time that may be a valid ground for objection to the nomination. As to the effect of the delays which may then follow, see the comments in the next chapter.

16.4 Design requirements

Clause 58(3) of the Sixth edition is identical to clause 58(3) in the

Fifth. It provides that design requirements are to be expressly stated for any work, goods or services ordered under a provisional sum or prime cost item. It further provides that design requirements shall be included in 'any' nominated sub-contract.

The wording of clause 58(3) is not fully clear and it is arguable that a design responsibility can only be placed on the contractor under a provisional sum or prime cost item when it is both expressly provided for in the contract and included in a nominated sub-contract. If that is the case, as some commentators suggest, then the engineer cannot order the contractor 'himself' to take on design responsibility under this clause even if it is expressly provided for in the contract.

Design obligations

There is also some uncertainty as to how the final sentence of clause 58(3) is to be interpreted. That sentence reads:

> 'The obligation of the Contractor in respect thereof shall only be that which has been expressly stated in accordance with this sub-clause.'

The question is: can a 'fitness for purpose' design obligation be imposed under clause 58 or does the limitation on responsibility for design in clause 8(2) (if indeed it is a limitation) apply to clause 58 and exclude fitness for purpose?

If fitness for purpose is so excluded it defeats much of the purpose of nominated sub-contracting which is frequently used to incorporate into the works specialist equipment selected precisely because of its fitness for purpose.

These complex issues have been debated since the publication of the Fifth edition and it is disappointing that they remain to trouble users of the Sixth edition.

Chapter 17

Nominated sub-contractors

17.1 Introduction

Clause 59 on nominated sub-contracting was the longest clause in the Fifth edition and it remains so in the Sixth edition.

The explanation for this is to be found in the words of Sir William Harris, who said in 1973, as chairman of the Joint Contracts Committee when replying to criticism of the newly introduced Fifth edition'

> 'These were undoubtedly the most difficult clauses to deal with, because the whole concept of nominated sub-contractors raises many complex difficulties, as recent decisions in the courts show. One of the most serious of these difficulties is the problem of the defaulting or insolvent sub-contractor who is a source of great potential loss and increased expense. The Committee had to decide upon whom that risk should rest: upon the employer who chose the sub-contractor, and ordered the contractor to employ him, or the contractor who was obliged to employ him. The Committee decided that, despite the power of objection given to the contractor, he who called the tune should pay the piper and the clauses are designed so that if loss should ultimately be suffered as a result of his chosen specialists's fault or bankruptcy, that loss should be upon the employer. The contractor, however, is under an obligation to do all he can to avoid such loss and to recover it from the sub-contractor. Where one has, as a basic premise, an artificial situation, it must follow that clauses designed to give it legal enforceability will themselves appear to be somewhat tortuous and complicated'.

The problem with indemnities

The basic question in nominated sub-contracting is how much

responsibility should the employer take for the defaults or failures of the sub-contractor he has selected and imposed on the contractor? That question was very much in the mind of the drafting committee of the Fifth edition because of the decision of the House of Lords in 1970 in the case of *North West Metropolitan Hospital Board* v. *T. A. Bickerton & Sons* (1970). The case illustrated the problem that if the employer indemnifies the contractor against loss caused by the nominated sub-contractor, the contractual chain of responsibility is broken since the contractor suffers no loss. The defaulting sub-contractor then escapes scot-free and the employer cannot recover his loss.

Policy of the Fifth edition

Standard forms of construction contracts issued since the *Bickerton* decision have approached the problem of indemnities with varying degrees of caution but this is how Sir William Harris described the policy for the Fifth edition.

> 'The Committee took the view that, as far as possible, the chain of liability should not be broken and that the responsibilities and liabilities should be spelt out clearly: also that employers and engineers should face up to this situation and consider carefully, in each case, whether "nomination" is in fact necessary.'

Policy of the Sixth edition

There has been no suggestion from the drafting committee of the Sixth edition that any major change of policy has been introduced on nominated sub-contracting. However, as will be seen later in this chapter some of the wording changes may have significantly increased the contractor's responsibility for a nominated sub-contractor's default.

The pitfalls of nomination

There is a warning in the words of Sir William Harris quoted above that employers and engineers should consider carefully whether nomination is necessary. Four good reasons can be given for this:

(a) the employer is taking financial risk which could be avoided by ordinary sub-contracting
(b) the contractual provisions are too complicated for lawyers to understand and agree upon, less still engineers
(c) too many 'what if' situations can be visualised to give any degree of certainty to the outcome of actions by the parties or the engineer
(d) the potential problems for the engineer and employer if the contractor objects to a nomination or seeks to terminate the sub-contract are immense.

There can be little doubt that unnecessary use of nomination by an engineer could leave him liable to the employer for negligence if things go wrong and the employer has to foot the bill under the contract.

17.2 General comparison – Fifth and Sixth editions

Clause 59 of the Fifth edition was in three parts:

59(A) – objections to nomination, responsibility, breach
59(B) – forfeiture/termination of the sub-contract
59(C) – payment to nominated sub-contractors

The Sixth edition incorporates all three parts into a single clause. The clearly intended major changes are:

(a) additional grounds for objection to nomination
(b) amendments to the engineer's action on objection to nomination
(c) amendments to the engineer's action in the event of termination
(d) termination without the employer's consent is no longer contemplated

A significant change, but one which may not have been intended since it clearly changes the balance of risk between the parties, is that the financial provisions for breach of the sub-contract now only apply in respect of termination.

17.3 Objection to nomination

Clause 59(1) of the Sixth edition details the contractor's rights to object to a nomination and clause 59(2) details the engineer's action upon objection. In the Fifth edition clauses 59(A) (1), (2) and (3) applied.

Under clause 59(1), the contractor is given the general ground of 'reasonable objection' and further grounds where the nominated sub-contractor declines to enter into a sub-contract containing provisions:

(a) that the sub-contractor will undertake such obligations and liabilities as will enable the contractor to discharge his own liabilities to the employer
(b) that the sub-contractor will indemnify the contractor against all claims arising out of failure in performing the obligations of the sub-contract
(c) that the sub-contractor will indemnify the contractor against claims for negligence
(d) that the sub-contractor will provide the contractor with security for performance of the sub-contract
(e) equivalent to those in clause 63 on determination.

Only category (d) on security for performance is new to the Sixth edition but it adds to what was, in any event, a formidable set of grounds for objection.

Reasonable objection

Under the Fifth edition concern on financial standing was generally thought to be within the scope of, and the likely main cause of, reasonable objection. However the separate requirement for security of performance in the Sixth edition will reduce the influence of financial standing under 'reasonable objections'. That leaves such issues as inadequate insurance, lack of competence, previous poor performance and commercial conflict as the obvious front runners.

Whether the knowledge that a particular nominated sub-contractor is claims-oriented can be the basis for reasonable objection is a moot point. It is difficult to see how the employer can complain if he has used 'a hard attitude to claims' as a disqualifying factor in his selection of the list of potential main contractors. And indeed questions on the main contractor's attitudes to claims usually figure on forms for references.

Like obligations and liabilities

The commonest problems with the sub-contractor declining to take on like obligations and liabilities as the contractor come in relation to time and liquidated damages. The contractor cannot object to a nominated sub-contractor who will not operate within the limits of a shortened programme but he has firm grounds for objection if the sub-contractor's work cannot be programmed to fit within the contract time. The employer's dilemma in such a situation was highlighted in the case of *Trollope & Colls* v. *North West Metropolitan Regional Hospital Board* (1973).

In that case the time remaining for phase III of a hospital building contract, after extensions granted on phases I and II, was 16 months instead of the 30 months originally intended. The employers, finding themselves unable to nominate sub-contractors for phase III who could complete in the shorter time, argued for an implied term in the contract that an extension should be granted to phase III to accommodate the delays in phases I and II. The contractors opposed the granting of any such extension. They were, in the words of Lord Pearson:

> 'turning the situation to their own advantage, because, if the contract could not be carried out, a new arrangement would have to be made for the work to be done at the prices prevailing in or about 1971, which were considerably higher than the contract prices. The difference between the contract prices and the prices prevailing in or about 1971 is said to be in the region of one million pounds.'

In a later building case *Fairclough Building Ltd* v. *Rhuddlan Borough Council* (1985) it was held that an architect's instruction in nominating a sub-contractor who could not complete within the time remaining after an earlier nominated sub-contractor had defaulted was invalid and the contractor was entitled to refuse the nomination.

Extending time to accommodate

When problems do arise in civil engineering contracts on

inadequate time for the nominated sub-contract work it is usual for the engineer to extend time but it is challengeable whether this can be done under the 'other special circumstances' of clause 44.

It is interesting to note that the special provision in clause 59B(4)(b) of the Fifth edition, allowing the engineer to extend the contract time under clause 44 for delay caused by forfeiture of a nominated sub-contract, is omitted from the Sixth edition.

Perhaps the best way for the engineer to get out of any dilemma on time is to give instructions under clause 13 or variations under clause 51. An extension can then be granted. It will cost the employer, of course, but that is the price of nominating sub-contractors.

Indemnity against claims

The indemnity against claims the contractor is entitled to under clause 59(1)(b) is not restricted to claims from the employer. It can and should include the contractor's own costs and claims from other sub-contractors.

Indemnity against negligence

The contractor would do well to check the nominated sub-contractor's insurances before accepting the nomination since an indemnity against negligence could be a worthless document if lacking financial support.

Security for performance

The new provision in clause 59(1)(d) for the sub-contractor to provide the contractor with security says nothing about the level of security or the form it should take.

Security under the main contract is regulated by clause 10 and is limited to 10% of the tender total. Can the contractor require a higher level than this from the nominated sub-contractor? Is there anything in clause 59 to say his requirement should be reasonable? Can the contractor require an on-demand bond rather than a default bond of the type in the main contract?

These are potentially contentious matters with dire consequences if they cannot be resolved.

Provisions equivalent to clause 63

The intention of clause 59(1)(e), which matches clause 59A(1)(d) in the Fifth, is not fully clear.

Clause 63 provides a detailed procedure for determination of the contractor's employment upon certain defaults. For the non-financial matters, the engineer plays a crucial role in its operation.

The question is should the 'equivalent' provisions include a role for the engineer, with the contractor substituting for just the employer; or should the contractor substitute for both the engineer and the employer? Given the wording of clause 59(4)(a) which says 'which in the opinion of the contractor justifies the exercise of his right', it would seem that the engineer is excluded from the procedure. That still leaves some uncertainty as to whether the 'opinion of the engineer' remains a constituent of the grounds for determination.

17.4 Engineer's action on objection (or upon determination)

Clause 59(2) brings together the actions the engineer shall take when the contractor objects to a nominated sub-contractor or there is a valid termination of the employment of a nominated sub-contractor. In the Fifth edition clauses 59A(2), 59A(3) and 59B(4) applied.

There are two significant changes from the Fifth edition. The engineer can no longer direct the contractor to enter into a nominated sub-contract (Fifth edition clause 59A(2)(c)) but there is a new provision in clause 59(2)(d) for the engineer to instruct the contractor to secure a sub-contractor of his own choice.

Courses of action

Under clause 59(2) the engineer has five possible courses of action:

(a) to nominate an alternative sub-contractor
(b) to vary the works under clause 51
(c) to omit the nominated sub-contract work by order under clause 51
(d) to instruct the contractor to secure a sub-contractor of his own choice

(e) to invite the contractor to execute the work himself

Each of these deserves further comment.

Re-nomination

The express power of the engineer to re-nominate covers the problem in the *Bickerton* case mentioned earlier where the employer maintained, after the first nominated sub-contractor had defaulted, that he was not bound to re-nominate or pay more than the original price.

The provisions for payment in clause 59(5) make clear that under these Conditions the actual price incurred must be paid.

Variation

If the engineer can overcome the problem of objection or termination by varying the works so as to avoid re-nominating he can do so under clause 59(2) (b).

Such a variation would be valued under the ordinary rules of clause 52 and it would not be a matter of meeting any price the contractor quoted as with valuation under clause 58(2) (b).

The contractor does not remain entitled to his percentage for charges and profit where there is a variation as he does for certain omissions (see below). In the Fifth edition, variations and omissions were grouped together in clause 59A(2) (b) and it was usually argued by contractors, with some success, that the words 'the omission of any Prime Cost item' applied both to variations and omissions.

This may be an unintended change or it may be intended as a matter of policy.

Omission

The express power in clause 59(2) (c) relates only to an omission where the work is to be carried out by the employer either concurrently or at some later date.

The power is given to avoid the argument that any omission of work necessary for completion is a breach of contract.

Such an omission cannot be ordered under clause 51 without

consideration of the payment of damages to the contractor and in recognition of this clause 59(2)(c) provides that the contractor is entitled to his charges and profit on the estimated value of the work omitted.

If the engineer can devise a way of omitting the work permanently, such an omission can be made under clause 59(2)(b). The contractor then has no entitlement to charges and profit.

Contractor's own choice

The intention of clause 59(2)(d) is presumably that the engineer can instruct the contractor to secure a 'nominated' sub-contractor of his own choice; that is, the employment of the sub-contractor would come within the provisions of clause 59 and not clause 4 (ordinary sub-contractors).

The point is not fully clear but the employment of ordinary sub-contractors would seem to come more naturally under clause 59(2)(e).

This new provision is, in theory at least, a sensible way of dealing with objections to nominations. The engineer is saying to the contractor – 'if you do not like my choice, make your own'.

For technical reasons it may not always work and there may be some misgivings on collusion, particularly if the re-nomination is at a higher price. But it does have the advantage of returning to the contractor some of the freedom of choice which is the basic characteristic of contracting.

Contractor to execute

The provision in clause 59(2)(e) for the engineer to invite the contractor to undertake the work himself is simply a return to provisions in clause 58 for the use of provisional sums and prime cost items.

It might be said that if this is a solution to a problem of objection to nomination or termination of the sub-contract, it is a solution which could, and should, have been used to avoid nomination in the first place.

But against this, there is the point that the work to be completed after termination of a sub-contract may be within the capability of the contractor and in such circumstances re-nomination would be unnecessary.

Delay in re-nomination

Comment has been made earlier in this chapter on the omission from the Sixth edition of the provision in the Fifth for extending time for delay consequent upon valid termination of a nominated sub-contractor's employment.

Wider questions on this issue are to what extent is the employer liable for delays in nomination, re-nomination or alternative action by the engineer; and what are the contractor's entitlements to extensions of time?

These were matters considered in the case of *Percy Bilton Ltd* v. *Greater London Council* (1982) when under a building contract the withdrawal of a nominated sub-contractor created direct delay, and failure to re-nominate promptly created further delay. It was held that the employer was responsible for only the second delay. The ruling dispelled the proposition that the *Bickerton* case had established automatic liability of the employer in the event of nominated sub-contractor withdrawal and it restated the general law that the contractor takes the risk unless there is fault by the employer. Lord Fraser said this:

> 'When the nominated sub-contractor withdrew, the duty of the employer, acting through his architect, was in my opinion limited to giving instructions for nomination of a replacement within a reasonable time after receiving a specific application in writing from the main contractor.'

Lord Fraser then went on to show how express terms of the contract could amend the general rule and give the contractor an entitlement to extension of time.

In the Sixth edition, the express terms are limited to an indemnity for cost which cannot be recovered from the sub-contractor. The intention is clearly that the contractor is responsible not only for the nominated sub-contractor's performance or lack of it, but also for his withdrawal. The Conditions say nothing on delay in nomination, re-nomination or other action. Such delay would have to be proved as breach of contract unless clause 13 could be shown to apply.

Clause 13 provides the contractor with the remedy for delay and cost he could have foreseen at the time of tender but he needs to have an engineer's instruction. The contractor therefore should not accept any nomination or re-nomination which comes too late to avoid delay in completion without first asking for an instruction under clause 13.

An alternative remedy might be found under clause 7 for delay in giving instructions but there is a difference in principle between the two clauses. The contractor's entitlement under clause 13 applies to any delay which could not have been foreseen; thus the promptitude of the engineer in dealing with matters affects only the length of the delay not the right of claim. Under clause 7, however, the right of claim arises only when there has been failure by the engineer to act within a time 'reasonable in all the circumstances'.

Implications of an unreasonable objection

Clause 59(1) entitles the contractor to raise a reasonable objection to a nominated sub-contractor. Clause 59(2) outlines the procedure when the contractor 'declines' to enter into a sub-contract.

Various commentators have suggested that an 'unreasonable' objection is a breach of contract and the contractor is responsible for any delay. This is no doubt true but it is a difficult matter to deal with in practice.

If the contractor holds to his objection, whether reasonable or not, and declines to enter into a sub-contract then clause 59(2) applies. There is no stated alternative course of action.

However, before embarking on unreasonable objections the contractor should note that his entitlement to payment under clause 13 is only to 'such cost as may be reasonable' thus unreasonable behaviour, if it could be proved, would lose the entitlement.

Design requirements

Comment has already been made on whether, under clause 58(3), design requirements in respect of provisional sums and prime cost items must be both expressly stated and included in a nominated sub-contract.

If both apply, then the engineer's scope for action under clause 59(2) is clearly limited where there is a design requirement. He can only re-nominate.

That is consistent with other provisions of the contract which indicate that design requirements must be expressly stated and cannot be imposed. An alternative solution for the engineer would be to seek the contractor's agreement to the contractor himself taking on the design responsibility.

17.5 Contractor responsible for nominated sub-contractors

Clause 59(3) repeats the provision in clause 59(4) of the Fifth edition that the contractor is responsible for the work, goods or services of a nominated sub-contractor as if he had executed or provided them himself.

This maintains the chain of contractual responsibility and it confirms the general legal rule, as stated in the *Bilton* case, that the contractor, not the employer is responsible for the nominated sub-contractor's defaults. However, some relief is given by clause 59(4) as shown below.

Contractors frequently find it hard to accept that they are not entitled to extensions of time when nominated sub-contractors have caused delay. But that is the case under the ICE Conditions, both Fifth and Sixth editions.

Contractors should note that for non-performing nominated sub-contractors the employer provides an indemnity on loss only if there is termination of the sub-contract. See also below the involvement of clause 13 following an application to terminate.

The contractor's best course of action therefore is to give notice of termination as rapidly as the contractual provisions allow.

Note that there is no equivalent of clause 59A(6) in the Sixth edition. That clause effectively provided an indemnity by the employer against the nominated sub-contractor's default. Now, as mentioned above, the indemnity applies only in the event of termination.

17.6 Nominated sub-contractor's default

Clause 59(4) covers exercise of the contractor's rights under the 'forfeiture' clause of the sub-contract. The reference to 'forfeiture' is a lapse back to the wording of the Fifth edition. The Sixth edition uses 'determination' in clause 63 and should have done so here.

Notice of default

Under clause 59(4) (a) if, in the opinion of the contractor, there are grounds for termination, the contractor shall 'at once' notify the engineer in writing. This is stated as a duty, so the contractor is bound to give notice whether or not he intends to proceed further.

The provision ensures that the engineer is alerted to a non-performing nominated sub-contractor sooner rather than later.

Consent to terminate

The contractor may only terminate the sub-contract with the consent in writing of the engineer – clause 59(4)(b). However, if the consent of the engineer is withheld, the contractor is entitled to instructions under clause 13.

Termination without consent

The provisions in clause 59B(5) of the Fifth edition for termination without consent are omitted from the Sixth edition.

If under the Sixth edition the contractor does terminate without consent that is a breach of contract entitling the employer to such damages as can be proved.

Instructions under clause 13

The effect of the provision in clause 59(4)(b) that, where the consent of the engineer is withheld the contractor is entitled to instructions under clause 13, may amount to an indemnity to the contractor against further losses arising from the nominated sub-contractor's defaults.

This may not have been the intention and it may be only practical measures which are envisaged as instructions under clause 13. But it is difficult to counter the argument that if the contractor applies to terminate the sub-contract and the engineer says 'no', then, as the clause is worded, the engineer is obliged to direct what happens next.

Engineer's action on termination

When the nominated sub-contractor's employment is terminated, clause 59(4)(c) requires the engineer to take action as set out in clause 59(2) for refusal. The implications of this have already been considered above.

Delay and extra expense

Clause 59(4) (d) is further maintenance of the chain of contractual responsibility. It places an obligation on the contractor to take 'all necessary steps and proceedings' available to recover the additional expenses of both the contractor and the employer.

Such a provision would have no place in the contract were it not for the indemnity given in clause 59(4) (e) by the employer to reimburse the contractor for all unrecoverable expenses.

'All necessary steps and proceedings' would include actions in arbitration or litigation as appropriate.

The reference in clause 59(4) (d) to recovering 'security' would require the contractor to call on any bond he had obtained from the nominated sub-contractor.

The additional expenses of the contractor could include his own costs of delay and disruption and claims from other sub-contractors plus legal costs. The additional expenses of the employer would most likely be payments to the contractor and payments to professional advisers.

The cost of delay is not directly addressed in clause 59(4) and there is no provision for an extension of time as there was in the Fifth edition. The intention is that no extension of time is granted and that liquidated damages for delay are payable. This may seem hard on the contractor but it is not really so since the contractor can recover the damages from the employer as a last resort under clause 59(4) (e).

This somewhat artificial arrangement is necessary to maintain the nominated sub-contractor's liability. In a building case, *Mellowes PPG Ltd* v. *Snelling Construction Ltd* (1989), the contract provided that an extension of time should be granted to the main contractor for delay caused by the sub-contractor. The contractor was unable to recover liquidated damages from the sub-contractor because, by his entitlement to an extension of time, he had suffered no loss himself.

Reimbursement of contractor's loss

The indemnity in clause 59(4) (e) that the employer will reimburse the contractor's unrecovered expenses following termination of a nominated sub-contract goes a long way towards relieving the contractor of the apparent burden of full responsibility imposed by clause 59(3).

However, the timescale for settling the amounts due is likely to be protracted. The contractor may have to wait to see what he can recover as an unsecured creditor when there has been a liquidation or he may have to sue in arbitration or in litigation. The contractor is unlikely to know at the date the submission of the final account becomes due what his position is and how much he can claim from the employer.

Fortunately the wording of clause 60(4) on the submission of the final account leaves an opening in referring to amounts due up to the date of the defects correction certificate.

17.7 *Provisions for payment*

The provisions for payment for nominated sub-contract work, goods or services in clause 59(5) are essentially the same as those in the Fifth edition.

The contractor is entitled to have included in the contract price:

(a) the actual price paid, or due to be paid, in accordance with the terms of the sub-contract net of all trade discounts
(b) the sum provided in the bill of quantities for labours
(c) a sum for charges and profit based on the rate in the bill of quantities or, in its absence, the rate in the appendix to the form of tender.

Default of the contractor

The phrase in brackets in clause 59(5)(a) relating to defaults is to ensure that the contractor cannot pass on to the employer any extra payments he is obliged to make on his own account to the nominated sub-contractor arising from his (the contractor's) defaults.

'Defaults' is possibly too narrow a word to be fully effective since the contractor may make other payments to the nominated sub-contractor which are neither defaults nor the responsibility of the employer.

Discounts

Clause 59(5)(a) reduces the price payable by the employer for trade discounts etc. but allows the contractor to retain the benefit of any discount he can obtain for prompt payment.

This has been criticised as open to abuse and prompt payment discount would certainly be a commercial bargaining factor if the contractor was free to make his own choice of sub-contractor.

Percentage for other charges and profit

Under clause 59(5) (c) the contractor is entitled to a percentage of the actual price at the rate in the bill of quantities or 'where no such provision has been made' at the rate in the appendix.

Engineers should note that while the percentage stated in the bill of quantities will be computed in the tender total, the percentage in the appendix will not.

Production of vouchers

Clause 59(6) provides, very sensibly in the light of employer's payment obligations, that the contractor shall produce to the engineer, when required, all quotations, invoices etc.

17.8 Direct payment to nominated sub-contractors

Clause 59(7), which is unchanged from the Fifth edition clause 59(c), fulfils two functions. Firstly, it empowers the engineer to check that payments are being properly made to a nominated sub-contractor, and secondly, it allows the employer to make direct payments if the contractor defaults in his payments.

The clause does not place any obligation on the employer to pay a nominated sub-contractor, even if there is default; nor does it create a right of action against the employer. It does, however, provide comfort to nominated sub-contractors and it is certainly of benefit if the main contractor becomes insolvent because the employer can pay directly for work which has been certified.

Broadly the provisions of clause 59(7) are:

(a) before issuing a payment certificate to the contractor under clause 60 the engineer is entitled to proof that all nominated sub-contractors have been paid amounts certified in previous certificates

(b) if any nominated sub-contractor has not been paid the contractor is required to give reasons and proof that he has informed the nominated sub-contractor in writing of his reasons for non-payment or set-off
(c) if the contractor fails to give such reasons and proof, the employer is entitled to pay the nominated sub-contractor the amount certified by the engineer which the contractor has failed to pay
(d) the employer can deduct any direct payments so made from amounts due to the contractor by way of set-off
(e) amounts so set-off by the employer shall be deducted from the amounts due on future certificates but the issue of such certificates shall not be delayed.

Note that under clause 60 amounts payable in respect of nominated sub-contractors are to be shown separately in the contractor's monthly statements and the certificates of the engineer.

Legal position on direct payments

It is usual for receivers or liquidators of an insolvent contractor to claim from the employer the full value of unpaid certificates and uncertified work in progress. That is because, as legal successors of the contractor, they have acquired his rights.

If the employer pays any sub-contractor, nominated or otherwise, for work, goods or services, undertaken or provided prior to receivership he will, unless special circumstances apply, remain liable to the receiver or liquidator for the same sum. He cannot discharge his legal debt to the receiver or liquidator by paying someone else.

The provisions in clause 59(7), if operated properly, protect the employer who has made direct payments since they are made under a contract which is itself still legally effective. However, the issues involved are unusually complex and justify taking legal advice before any direct payments are made.

Chapter 18

Certificates and payments

18.1 *Introduction*

This chapter examines clause 60 on monthly payments, the final account, retentions, interest on overdue payments and the correction and withholding of certificates.

The changes from the Fifth edition are mainly procedural or points of clarification but the expanded clause 60(7) on interest on overdue payments may have a significant impact on the attitude of the parties and the approach of the engineer to certification.

18.2 *Contractor's monthly statements*

The requirement in clause 60(1) for the contractor to submit monthly statements has been changed from 'after the end of each month' in the Fifth edition to 'at monthly intervals' in the Sixth.

The statement is to show:

(a) the estimated contract value of the permanent works executed up to the end of that month
(b) a list of goods or materials delivered to site but not incorporated in the permanent works and their value
(c) a list of goods or materials identified in the appendix but not delivered to site property in which has vested in the employer and their value
(d) the estimated amounts the contractor considers himself entitled to under the contract for other matters including temporary works.

A monthly statement need not be submitted if, in the opinion of the contractor, the amount will not justify the issue of an interim certificate.

Amounts payable in respect of nominated sub-contractors are to be listed separately.

Value of permanent works

The essence of the scheme for payments is that each statement should show cumulative value and not incremental adjustment of the previous statement. Where the engineer has prescribed a form in the specification, as contemplated by clause 60(1), the statement will invariably be submitted in cumulative form. Where the contractor is left with freedom to devise his own form he may slip into habits used elsewhere in the construction industry of incremental statements.

It is suggested that the phrase 'estimated contract value' does not relieve the contractor of presenting a measured account each month. The Sixth edition, like the Fifth, is a measure and value contract and whilst stage payments, activity schedules and percentage completions may be appropriate for lump sum contracts that is only because the final sum payable is broadly known at the outset.

Clause 60(1) (a) retains the phrase 'up to the end of that month' although the requirement for statements has been changed to 'at monthly intervals'. Perhaps the wording could have been clearer. It is not thought that the value should be to the end of a calendar month.

Goods on site

The only certainty about the ownership of goods on site is that, when they are fixed on the employer's land, he has good title. For the rest, whatever contracts of sale or construction contracts may say, possession remains the strongest point from which to argue. Since average payment times on invoices are, in the UK, in excess of two months it follows that more often than not the contractor is paid for materials on site before the supplier is paid. Not surprisingly, suppliers often turn up on site and take their goods away when a contractor goes out of business.

Security of the site should, therefore, be an important consideration in the payment for unfixed goods and materials – see the comment in chapter 2 on the dangers of extending the site.

Goods off-site

Payment for goods off-site is even more risky than payment for goods on-site. Receivers and liquidators are not readily persuaded that things within the curtilage they control are not within their power to dispose of. Clause 54 lays down an elaborate procedure for vesting goods and materials not on site in the employer but it often proves to be of no consequence. See the discussion in section 2 of chapter 14 under 'Retention of title'.

If the employer is prepared to take the risk of paying for off-site goods, the key point for the engineer to establish is ownership. Payment to the contractor for goods he does not own can only give the employer claim to title in exceptional circumstances.

Value of goods and materials

Unfixed goods and materials may have two widely differing values – that as invoiced, and that as marked-up in the bill of quantities. The basis of payment until they are fixed should be as invoiced and net of discounts.

Other matters

The provision in clause 60(1) (d) for inclusion in the contractor's statement of the estimated amounts in connection with 'other matters' covers most obviously the claim clauses of the contract. Thus clauses 13(3), 14(8) and others refer expressly to payment in accordance with clause 60. Clause 52(4) (f) on claims generally confirms that the contractor is entitled to have included in any interim payment certified under clause 60 such amount as the engineer considers due.

Other clauses which can give the contractor an entitlement to payment but make no express reference to clause 60, such as clauses 17 (setting-out) and 38 (uncovering), will also be within the scope of 'other matters' in clause 60.

Inclusion of claims

The provision that the contractor 'shall' submit at monthly intervals a statement showing the estimated amounts to which he considers himself entitled is sometimes used as the basis for rejecting a contractor's claim which is lodged late or post-completion.

The provision needs to be read in conjunction with clause 52(4) and it is difficult to see how failure by the contractor to include a claim does any more than deprive him of his right to interim payment.

Extra-contractual claims

It is doubtful if claims for breach of contract, not expressly covered in the Conditions, should be included as 'other matters' within clause 60. Such claims could include failures by the engineer to perform his duties or by the employer to fulfil his obligations.

The wording in clause 60(1)(d) – 'for which provision is made under the Contract' – certainly indicates that only contractual claims are covered by the clause and this suggests that submissions and payments for breach are to be dealt with elsewhere. There is, however, a complication.

Although a contract administrator would not normally have power to adjudge and certify on extra-contractual claims, the engineer in the ICE Conditions is under a duty by clause 66 to decide on all disputes referred to him 'in connection with or arising out of the contract'. This clearly includes extra-contractual claims. If the engineer finds in favour of the contractor, can it not be said that payment is due because provision has been 'made under the Contract'?

Retention on claims

A small financial point against including extra-contractual claims in the clause 60 procedure is that retention is deducted under 60(5) from the amount due under 60(1)(d).

In principle, retention should not be deducted from any claim payment which is by way of payment for breach but insofar that deduction is allowed under clause 60(5) it must be accepted. However, settlement of an extra-contractual claim, whether under clause 66 or by agreement, should be free of retention.

Value added tax

There is a further complication on value added tax as to whether it applies equally to the provisions of goods and services and damages for breach of contract when payments are made together under the contract. Court awards for goods and services attract VAT; court awards for damages do not. Resolution of the position in contracts is a matter for experts.

Interest on late payment

By clause 60(7) the contractor is entitled to interest on late payments. The wording of the clause can be taken as an obligation on the employer to pay interest whether or not there is an application from the contractor.

In practice, for reasons discussed later in this chapter, contractors will usually, as a matter of prudence, include any interest they consider due under clause 60(7) in their clause 60(1) statements.

18.3 Monthly payments

Clause 60(2) provides that, within 28 days of the date of delivery of the contractor's monthly statement, the engineer shall certify and the employer shall pay the amount due.

This arrangement is open to the criticism that the engineer, by certifying late in the period, can leave the employer with very little time to arrange his funds. Nevertheless, it seems to work reasonably well.

Non-payment of certificates

Unlike most standard forms of construction contract, the ICE Conditions do not give the contractor the express right to determine his own employment for non-payment of a certificate. The only express remedy provided is the payment of interest under clause 60(7).

Repeated failure to pay on certificates could entitle the contractor to exercise his common law rights of determination. See also the case of *Fernbrook Trading*, mentioned in chapter 9, on the contractor's claims for delay caused by late payments.

Order 14 applications

Order 14 of the Rules of the Supreme Court provides a procedure for a claimant to obtain summary judgment where there is no arguable defence.

A contractor faced with non-payment of a certificate would probably be advised by his lawyers that the employer was in debt, not in dispute and that an application to the courts under Order 14 would be a speedier way of obtaining payment than use of the contractual dispute procedure.

The success of such an application would depend very much on the response of the employer and the case he could put up for non-payment.

Employer's set-off

One obvious defence for the employer to an Order 14 application is that he has a counterclaim which he is entitled to set off against the amount certified by the engineer.

The general right of an employer to counterclaim and set-off against amounts certified was re-affirmed in *Gilbert-Ash (Northern) Ltd* v. *Modern Engineering (Bristol) Ltd* (1973) following uncertainty caused by the decision in *Dawnays Ltd* v. *Minter* (1971). There can, of course, be contractual provisions, to exclude or modify the general right.

The position on the ICE Conditions was clarified in the recent case of *Enco Civil Engineering Ltd* v. *Zeus International Development Ltd* (1991). Disputes arose on the quality of work executed and the employer failed to pay on two certificates. When the contractor applied for judgment under Order 14 for the amounts certified, the employer issued a summons to stay proceedings on the basis that disputes existed which had been referred to the engineer under clause 66. The court granted the stay and held additionally that there was nothing in the ICE Conditions which prevented the employer from setting-off any cross-claim against the amounts certified for alleged defects.

This case undermines the view that the employer is required to pay in full on each and every certificate and that he must raise any challenge through the disputes procedure of the contract, commencing with the engineer's clause 66 decision.

Date of delivery

It is important that the contractor can establish a firm date of delivery of his statement. In the *Enco Civil Engineering* case the contractor's application for judgment under Order 14 failed in any event in respect of one certificate because he was unable to establish when his statement had been delivered to the engineer.

The case highlighted a point which had, perhaps, been previously overlooked – that the engineer's certificate by itself does not establish any time for payment by the employer.

Amounts due

Clauses 60(2) (a) and 60(2) (b) provide differently worded tests for the amount due to the contractor. For work and claims it is the amount 'which in the opinion of the Engineer on the basis of the monthly statements is due', less retention. For unfixed goods and materials it is the amount 'the engineer may consider proper'.

Prior to the case of *Morgan Grenfell (Local Authority Finance) Ltd* v. *Seven Seas Dredging Ltd* (1990), discussed in detail later in this chapter under clause 60(7), the differences, as they stood then in the Fifth edition, appeared academic. But what is due 'in the opinion of the Engineer' may no longer be what is contractually due to the contractor; perhaps what the engineer considers 'proper' is still the amount which remains due.

As to the meaning of 'proper' perhaps the engineer could consider amongst other things:

(a) the security point mentioned above
(b) the condition of the goods
(c) the timing of their arrival on site
(d) proof of payment/ownership.

In practice if the engineer and contractor are in disagreement on the matter, the contractor can frequently do little other than complain since the timescale of his remedies (if any) would normally exceed the duration of the problem.

18.4 Minimum amount of certificates

Clause 60(3) provides that until substantial completion of the whole of the works the engineer is not bound to issue an interim certificate for a sum less than the minimum amount stated in the appendix.

This provision varies from that in the Fifth edition by making it clear that after the issue of the certificate of substantial completion for the whole of the works, the minimum limits no longer apply and payments are due at monthly intervals.

18.5 Final account

Clause 60(4) endeavours to regulate the timely completion of accounts by requiring the contractor to submit a statement of final account and supporting documentation not later than three months after the date of the defects correction certificate. The statement is to show the value of the works executed together with all further sums which the contractor considers due.

By clause 61(1) there is stated to be only one defects correction certificate and that follows expiration of the last defects correction period.

Failure to include sums due

Failure by the contractor to include sums due under the contract in his final account may deprive him of entitlement to payment since it is no longer a matter of securing interim payment. The contractor would, of course, retain his right to sue for damages for breach where that was an alternative.

Engineer's final certificate

Within three months after receipt of the final account and verifying information the engineer is required to issue a certificate stating the amount which in his opinion is finally due to either the contractor or the employer as the case may be.

Clause 60(4) does not use the phrase 'final certificate' and there is no status of finality attached to the certificate as there often is in building contracts where the final certificate is binding unless challenged within a specified time. Moreover since the 'final' certificate covers only sums considered due up to the date of the defects correction certificate, there may well be later certificates as disputes are settled or insurance related matters are resolved.

Engineer not *functus officio*

There is certainly nothing in clause 60(4) to suggest that the engineer is rendered *functus officio* after the issue of his 'final' certificate and indeed he may still have duties to perform under the contract such as giving decisions under clause 66 which might lead him to revising his final certificate.

Credits to the employer

The wording of clause 60(4) leads to some interesting questions on how the engineer should handle credits due to the employer.

The clause says that the engineer is to certify after giving credit to the employer 'for all sums to which the Employer is entitled under the contract'. It then says that the amount certified shall 'subject to clause 47 be paid'. That suggests the amount to be paid is the amount certified after deduction of any liquidated damages for late completion.

This indicates that the engineer should not deduct in this certificate, and certainly not in earlier certificates, for liquidated damages. This would be consistent with the wording of clause 47(5).

Employer's set-off

The employer has, of course, the same rights of set-off under clause 60(4) as he has under clause 60(2).

18.6 Retention

Clause 60(5) fixes the retention held under clause 60(2) (a) as the difference between:

(a) an amount calculated in and up to the limit in the appendix, and
(b) any payment which shall have become due under clause 60(6).

Clause 60(6) covers the release of retention which is due at its simplest in two stages:

(a) one half upon the issue of any certificate of substantial completion, and
(b) the balance upon the expiry of the final defects correction period.

Both amounts so due are to be paid within 14 days.

Where there are certificates of completion for sections or parts of the works, the first tranche of retention is released in proportion of the value of work completed to the whole, with the proviso that the amounts so released shall not exceed one half of the limit of retention set out in the appendix. This ensures that the employer has half the retention at substantial completion for the whole of the works.

The second tranche is only released upon the expiry of the last defects correction period and not in instalments where there is more than one such period.

Retention limits

The retention provisions of the Fifth edition in clause 60(4) stated limits of 5% up to £50,000 and 3% thereafter. In the Sixth edition the limits are stated in the appendix not in clause 60(5). The recommendation in the appendix is a limit not exceeding 3%.

Outstanding claims

Clause 60(6) (c) expressly states that the last tranche of retention money is to be paid upon the expiry of the last defects correction period notwithstanding any outstanding claims by the contractor against the employer.

Anything else would be patently unjust.

Outstanding work

Since it is the expiry of the defects correction period and not the issue of the defects correction certificate which triggers release of the second half of retention money, clause 60(6) sensibly provides that the employer can withhold payment of so much of the retention money as in the opinion of the engineer represents the cost of completing any outstanding work.

Retention in trust

The great debate in the building industry on the duty of employers to treat retention monies as held in trust has had little obvious impact in civil engineering.

There are significant differences in the drafting of the various forms of contract, most notably that the ICE Conditions do not expressly provide for the employer to hold retention as a trustee. Nevertheless it is likely that the pressures in the building industry for all retentions to be held in special accounts will at some time in the not too distant future spill over into civil engineering.

18.7 Interest on overdue payments

Clause 60(7) providing for interest on overdue payments is a modification of clause 60(6) in the Fifth edition.

The basic premise remains unaltered; that in the event of failure by the engineer to certify or the employer to make payment of the amounts due under clause 60 the employer shall pay interest at 2% above base lending rate.

The modifications, or clarifications as they have been called, relate to the detail of interest calculations and the meaning of 'failure to certify'.

Interest calculations

The scheme in clause 60(7) for the calculation of interest can be summarised as follows:

(a) Interest is to be compounded monthly for each day on which payment is overdue.
(b) Payment is overdue from the date it should have been paid (if certified) or the date it should have been certified and paid (if not certified).
(c) The interest rate is 2% above the base lending rate of the bank specified in the appendix.

(d) Interest will be added each month on interest already due to be paid.

Payment of interest

Clause 60(7) makes no provision for the engineer to certify interest on overdue payments. It is the responsibility of the employer to pay interest directly it becomes due.

However, the contractor would be wise to include any interest due in his monthly application under clause 60(1)(d) as being an amount for which provision is made under the contract. The engineer would then be obliged to certify such interest under clause 60(2)(a).

Findings of an arbitrator

Completely new to the Sixth edition is the provision in clause 60(7) that if an arbitrator finds that a sum should have been certified by a particular date but was not so certified, it shall be regarded as overdue for payment from either:

(a) 28 days after the date the arbitrator holds the engineer should have certified, or
(b) from the date of the certificate of substantial completion of the whole of the works where no certifying date is identified by the arbitrator.

This new provision is consistent with, and follows, the judgment in the case of *Morgan Grenfell (Local Authority Finance) Ltd* v. *Seven Seas Dredging* (1990).

In that case under the ICE Fifth edition Conditions the arbitrator found in favour of the contractor for unforeseen work in dredging for the Port of Sunderland. The contractor was awarded £1,954,811 plus interest. The arbitrator awarded interest of £967,604 calculated at 2% above bank rate under clause 60(6). If the arbitrator had calculated interest under section 19(A) of the Arbitration Act 1950 the award of interest would have been £187,449 less.

The employer appealed and the court had to decide whether a contractor is entitled to interest under clause 60(6) in respect of amounts included in a statement under clause 60(1) but not certified by the engineer, when an arbitrator later decides they should have been certified. The appeal was dismissed with the court finding that interest under clause 60(6) was allowable.

Meaning of failure to certify

At the heart of the *Morgan Grenfell* case was the long argued point of whether an engineer, in rejecting all or part of a contractor's statement, leaves the employer open to liability for interest if he (the engineer) or an arbitrator subsequently revises upwards the evaluation of the contractor's claim. A common view was that, providing the engineer acted in good faith in forming 'his opinion' under clause 60(2), there was no 'failure to certify'. This is how Judge Newey QC put the arguments of the parties in the *Morgan Grenfell* case:

> 'Counsel for the employer submitted to me that clause 60(2) required the engineer first to form an opinion as to the amount shown in the contractor's statement which was due to the contractor and then to issue a certificate as to that amount. If the engineer failed to form an opinion or either failed to issue a certificate or omitted a sum from it interest is payable under clause 60(6). If, on the other hand, the engineer formed an opinion and as a result decided not to certify or not to include an amount in a certificate, there has not been a 'failure to certify' and interest is not payable under clause 60(6).
>
> He said that withholding a certificate only amounts to failure to certify when the engineer has not acted *bona fide*.
>
> Counsel for the contractor contended that clause 60(6) applied when the engineer failed to certify sums properly due under clause 60(2). If the contractor submitted a statement under clause 60(1) and the engineer disallowed a sum properly due when he issued his certificate, he failed to certify in accordance with clause 60(2). Certificates issued by the engineer could be reviewed by himself under clause 60(7) or by an arbitrator under clause 66(1). If either the engineer or the arbitrator found that the engineer had wrongfully failed to issue a certificate or failed to certify the proper amount, then the error is corrected and the employer must under clause 60(6) pay interest from the date when the money was properly due.'

This is how the judge explained his findings:

'I have reached the same conclusions as the learned arbitrator. In my judgment clause 60(2) requires the engineer, who has been provided with a statement in accordance with clause 60(1), to "certify" the amount which the contractor should be paid. Obviously, that amount should be the amount, which in his "opinion" the contractor should be paid. If the word "opinion" had not been used in the sub-clause some similar word or words would have had to have been used. For the sub-clause to work the engineer has to decide how much the contractor should receive and that is a matter of "opinion", "judgment", "conclusion" or the like. If the engineer certifies an amount which is less than it should have been, the contractor is deprived of money on which he could have earned money.'

Implications of failing to certify

Now that the *Morgan Grenfell* decision is incorporated into the Sixth edition there may well be some encouragement to contractors to claim early, claim high and keep claiming interest. Engineers on the other hand will have to exercise greater caution in reaching their opinion on what is due. Summary rejection of a contractor's claim might later be viewed by the employer as negligence. The best course for engineers is to use to the full the provisions of clauses 52(4) and 60(4) on requiring substantiation of all amounts claimed.

Extra-contractual claims

It is worth noting that the interest provisions of clause 60(7) apply only to amounts due under the contract. The contractor therefore has no contractual entitlement to interest for late certification or payment of an extra-contractual claim.

Interest not financing charges

The interest in clause 60(7) should not be confused with the finance charge in 'cost' as defined in clause 1(5). Interest applies to late payment on a contractor's application; finance charges are part of the cost included in such an application.

18.8 Correction and withholding of certificates

Clause 60(8) empowers the engineer to omit from any certificate the value of work done or goods supplied with which he is dissatisfied. For that purpose, and for any other reason which 'may seem proper' the engineer may, in any certificate, delete, correct or modify any previous sum certified.

Dissatisfaction

The provision on dissatisfaction recognises that as the works proceed matters may come to light which diminish or possibly eliminate the value of work already paid for in that reconstruction or demolition may be necessary at the contractor's own cost.

Other reasons

There is some uncertainty on what is meant by 'any other reason which to him may seem proper'.

It is not thought to be a general power for the engineer to coerce or punish the contractor for contractual failings and it is most unlikely that it could apply to matters outside the contract – such as knowledge of the contractor's financial difficulties.

Possibly it relates to the withholding, particularly from interim certificates, of amounts due under the contract to the employer from the contractor. For example, in the case of clause 39(2) where the employer has carried out corrective work at the contractor's expense. However, that clause, and others similar, say the employer may deduct 'from any monies due' which suggests that the deduction should be by way of set-off and not by reduced certification.

Perhaps the provision means not a great deal more than the engineer can correct mistakes or change his mind.

Application to nominated sub-contractors

In respect of payments to nominated sub-contractors, the engineer's powers under clause 60(8) are restricted.

By clause 60(8)(a) the engineer may not reduce in an interim certificate the sum previously certified for a nominated sub-contractor if the contractor has already paid over that sum.

By clause 60(8) (b) if the engineer, in the final certificate, reduces the sum previously certified to a nominated sub-contractor and that sum has already been paid by the contractor, then the employer is obliged to reimburse the contractor for the amount overpaid to the extent that the contractor is unable to recover the same from the sub-contractor.

Numerous complex scenarios can be advanced whereby, under these provisions, the employer comes out the unfortunate and innocent loser. All that can be said, once again, is that this is the price of nominated sub-contracting.

18.9 Copy certificates

Clause 60(9) provides simply that every certificate issued by the engineer pursuant to clause 60 shall be sent to the employer and at the same time copied to the contractor 'with such detailed explanation as may be necessary'.

In practice it is usual for the engineer to return to the contractor a copy of the application for payment showing any reductions or amendments.

18.10 Payment advice

Clause 60(10) is a new provision which applies when the employer pays an amount different from that certified by the engineer.

The employer is to notify the contractor 'forthwith' with full details showing how the amount paid has been calculated.

This is a useful addition and a necessary one in the light of the changes to clause 47 on liquidated damages where the conditions precedent to deduction have been eliminated.

Chapter 19

Determination and frustration

19.1 *Introduction*

This chapter deals with the following clauses:

clause 63 – determination of the contractor's employment
clause 64 – frustration
clause 65 – war clause.

Clauses 64 and 65 are unchanged from the Fifth edition but clause 63 has been renamed and modified to take account of the freedom of the contractor to sub-contract without obtaining the consent of the engineer. A few drafting improvements have also been made.

19.2 *Determination generally*

The ordinary remedy for breach of contract is damages but there are circumstances in which the breach not only gives a right to damages but also entitles the innocent party to consider himself discharged from further performance.

Repudiation

Repudiation is an act or omission by one party which indicates that he does not intend to fulfil his obligations under the contract. In civil engineering works, a contractor who abandons the site or an employer who refuses to give possession of the site would be obvious examples of a party in repudiation.

Determination at common law

When there has been repudiation or a serious breach which goes to the heart of the contract, sometimes called 'fundamental' breach, common law allows the innocent party to accept the repudiation or the fundamental breach as grounds for determination of the contract. Practical effects of this could include a contractor leaving the site because of persistent non-payment or the employer expelling the contractor from the site for non-performance. The innocent party would then normally sue for damages on the contract which had been determined.

The problems with common law determination are that it is valid only in extreme circumstances and it can so readily be challenged that it lacks certainty. The danger for a party who, considering himself innocent and the other party at fault, determines the contract, is that if in a subsequent action his grounds for determination are found to be inadequate, he may then himself be judged to be the party in serious breach for determining.

Determination under contractual provisions

To extend and clarify the circumstances under which determination can validly be made and to regulate the procedures to be adopted, most standard forms of construction contract include provisions for determination. Building forms usually give express rights of determination to both employer and contractor but civil engineering forms commonly give express rights only to the employer.

Many of the grounds for determination given in standard forms are not effective for determination at common law. Thus failure by contractor to proceed with due diligence and failure to remove defective work are often to be found in contracts as grounds for determination by the employer. But at common law neither of these will ordinarily be a breach of contract at all since the contractor's obligation is only to finish on time and to have the finished work in satisfactory condition by that time.

The commonest and the most widely used express provisions for determination relate to financial failures or difficulties encountered by the contractor. Again at common law many of these are ineffective and even as express provisions they are often challenged as ineffective by legal successors of failed companies.

The very fact that grounds for determination under contractual provisions are wider than at common law leads to its own difficulties. A party is more likely to embark on a course of action when he sees his rights expressly stated than when he has to rely on common law rights and this itself can be an encouragement to error. Some of the best known legal cases on determination concern determinations made under express provisions but found on the facts to be lacking in validity.

In *Lubenham Fidelities* v. *South Pembrokeshire District Council* (1986) the contractor determined for alleged non-payment whilst the employer concurrently determined for failure to proceed regularly and diligently. On the facts, the contractor's determination was held to be invalid. But in *Hill & Sons Ltd* v. *London Borough of Camden* (1980), with a similar scenario, it was held on the facts that the contractor had validly determined.

Determination of the contractor's employment

Provisions for determination in construction contracts are usually drafted as provisions for determination of the contractor's employment.

The purpose of this is to emphasise that the contract itself remains in existence and that its various secondary provisions remain operative after determination – not least those in the determination clause itself.

Parallel rights of determination

Some construction contracts expressly state that their provisions, including those of determination, are without prejudice to any other rights the parties may possess. That is, the parties have parallel rights – those under the contract and those at common law – and they may elect to use either.

The ICE Conditions do not have such a stated alternative but the omission is not significant. The general rule is that common law rights can only be excluded by express terms. Contractual provisions, even though comprehensively drafted, do not imply exclusion of common law rights.

The point came up in the case of *Architectural Installation Services Ltd* v. *James Gibbons Windows Ltd* (1989) where it was held that a notice of determination did not validly meet the timing requirements of the contractual provisions but nevertheless there had been a valid determination at common law. In considering the argument that common law rights were excluded because the determination clause was not prefaced with the words 'without prejudice to other rights and remedies' Judge Bowsher QC said:

> 'I am not impressed by that submission and I would be sorry if draftsmen of contracts felt it necessary to include such legal verbiage in order to avoid unintended results of their drafting.'

He went on to say:

> 'I see no reason at all for any implication of the term to the effect that condition 8 of the contract is to be the only machinery for terminating the contract, to the exclusion of common law rights of termination.'

Legal alternatives

To take advantage of the parallel rights of determination and to avoid as far as possible the danger that one course of action might be found to be defective, determination notices are sometimes served under both the contract and at common law as legal alternatives.

But on that it must be said that contractual rights which are not also common law rights attract only contractual remedies. So a defective notice under a contractual provision which has no common law equivalent cannot be salvaged by a common law action.

Determination – a legal minefield

It will be clear from what has been said generally, and what follows specifically on the ICE Conditions, that determination is a legal minefield. One false move and the apparently innocent party is cast as the wrongdoer.

It is difficult to envisage any circumstances so clearcut that the consultation of lawyers can be neglected before determination action is taken. Expensive though lawyers may seem, their costs are nothing compared to those of a wrongful determination.

19.3 Determination under the ICE Conditions

The Fifth edition of ICE Conditions used the word 'forfeiture' as the marginal note to clause 63. The Sixth edition uses the phrase 'determination of the contractor's employment'.

Although clause 1 of the Conditions says that headings and marginal notes are not part of the contract and are not to be used in its interpretation, the change in the marginal note may be of some assistance in better identifying the purpose of clause 63.

The essence of clause 63 in both the Fifth and the Sixth editions is that the employer can expel the contractor from the site and then complete the works as he thinks fit. The Fifth edition used the phrase 'expel' throughout and never mentioned 'determination' or even 'forfeiture' in the text of clause 63. It did, however, refer to determination in clause 64 (frustration) and clause 65 (war clause). This led to the contention that clause 63 of the Fifth edition was not, in law, a determination provision.

In the case of *Dyer Ltd* v. *Simon Byld/Peter Lind Partnership* (1982) a sub-contract under the FCEC form (the 'blue form') entitled the main contractor to determine the sub-contract if the main contract was determined. The main contract was 'determined' under provisions the same as those in clause 63 of the Fifth edition. The judge upheld an arbitrator's finding that the expulsion of the contractor from the site under clause 63 did not determine the contract. Mr Justice Nolan said:

> 'I do not see how the invocation by the employer of his rights under clause 63 and his exercise of the continuing powers and acceptance of the continuing duties for which that clause provides can be said to have determined or terminated the contract.'

The decision in that case concerned primarily the interpretation of clause 16 of the 'blue form' of sub-contract but there is no certainty that the Sixth edition is in any different position than the Fifth in its linkage to the sub-contract. However it does, at least, include the phrase 'a notice of determination' in its text and might, therefore, be said to be more obviously a determination clause.

19.4 Grounds under clause 63

The grounds for determination of the contractor's employment under clause 63 fall into two categories:

(a) financial failures, clause 63(1)(a), and
(b) failures of performance, clause 63(1)(b).

Financial failures

Under clause 63(1) (a) (i) the contractor is in default for any of the following:

(a) becoming bankrupt
(b) having a receiving order or administration order made against him
(c) making an arrangement or assignment in favour of creditors
(d) agreeing to a committee of inspection
(e) going into liquidation.

Under clause 63(1) (a) (ii) the contractor is in default if he assigns the contract without consent or has an execution levied on his goods which is not stayed or discharged within 28 days.

Changes from the Fifth edition are the inclusion of an administration order as a default and the 28 day period of grace on an execution order.

Whilst the provisions may well be effective against the contractor himself it is questionable how effective they are against trustees in bankruptcy, liquidators, administrators and other legal successors of the contractor.

To some extent the provisions are perverse in that a contractor who has applied for administration or a creditors' voluntary arrangement is showing by his conduct that, far from repudiating his obligations under the contract, he is taking legal steps to enable him to complete them.

Failures of performance

Under clause 63(1) (b) the specified defaults are:

(a) abandoning the contract
(b) failing to commence without reasonable excuse
(c) suspending progress for 14 days after notice to proceed
(d) failing to remove or replace defective work, goods or materials after notice to do so
(e) failing to proceed with due diligence despite previous warnings
(f) being persistently or fundamentally in breach of obligations under the contract.

These are the same as in the Fifth edition but with the defaults of sub-letting to the detriment of good workmanship or in defiance of the engineer's instructions omitted.

'Abandoned the Contract'

Abandonment may be patently obvious where the contractor has left the site or has otherwise given notice of his intention not to fulfil his obligations. Such abandonment is repudiation and grounds for common law determination.

However, there can also be situations where the contractor's conduct in ceasing work is neither abandonment or repudiation. In *Hill* v. *Camden* (1981), the contractor, unable to obtain prompt payment cut his staff and labour. The employer took this action as repudiatory and served notice of determination. Lord Justice Lawton had this to say of the contractor's action:

> 'The plaintiffs did not abandon the site at all; they maintained on it their supervisory staff and they did nothing to encourage the nominated sub-contractors to leave. They also maintained the arrangements which they had previously made for the provision of canteen facilities and proper insurance cover for those working on the site.'

'Failed to commence'

By clause 41(2) the contractor shall start the works on or as soon as is reasonably practicable after the works commencement date. The default under clause 63 is failing to commence in accordance with clause 41 without reasonable excuse.

The manner in which the date is set might have some relevance to what is 'reasonably practicable' and what is a 'reasonable excuse'. Thus if the works commencement date specified in the appendix becomes coincident with the award of the contract or, where the date is to be notified within 28 days of award, it is notified on the last day, the contractor must be allowed some time to mobilise.

Failure to commence within a reasonable time is not of itself a repudiatory breach. The contractor might say that he can complete in a fraction of the time allowed and has every intention of finishing on time.

On a practical point, can it be said that a contractor who has arrived on site but is taking longer than expected to set up his compound or set-out the works has failed to commence? Does failure to commence imply total absence from the site and, if not, what level of presence and activity is needed as evidence of commencement?

Suspension of progress

It is not clear whether the default of suspending progress relates only to inactivity after a clause 40 suspension or whether it has some wider application. Nor is it clear if the words 'without reasonable excuse' apply to this default as well as to failure to commence.

The words 'after receiving from the engineer written notice to proceed' suggest that the suspension is related to some procedure in the contract, most obviously, an ordered suspension under clause 40. If the provision has wider application, the notice to proceed is an administrative requirement of clause 63 which is additional to other notices.

Failure to remove defective work, goods or materials

Failure by the contractor to remove or replace defective work, goods or materials is not, of itself, a breach of contract. It is only a breach of contract if the contractor has failed to do so by the completion date. In *Kaye Ltd* v. *Hosier & Dickinson Ltd* (1972) Lord Diplock said:

> 'During the construction period it may, and generally will, occur that from time to time some part of the works done by the contractor does not initially conform with the terms of the contract either because it is not in accordance with the contract drawings or the contract bills or because the quality of the workmanship or materials is below the standard required by condition 6(1).... Upon a legalistic analysis it might be argued that the temporary disconformity of any part of the works with the requirements of the contract, even though remedied before the end of the agreed construction period, constituted a breach of contract for which nominal damages would be recoverable. I do not think that makes business sense. Provided that the

> contractor puts it right timeously I do not think that the parties intended that any temporary disconformity should of itself amount to a breach of contract by the contractor.'

The ICE Conditions do, however, in clause 39, empower the engineer to order the removal of unsatisfactory work and materials and failure to comply with such an instruction is a breach of contract. The remedy of determination in clause 63 is extreme and not surprisingly there is little evidence of its use for this purpose. Perhaps the mere serving of the engineer's clause 63 notice is effective in itself in getting action from the contractor.

Prior to the case of *Tara Civil Engineering* v. *Moorfield Developments Ltd* (1989) it was generally thought that the engineer could not serve notice under clause 63 until he had acted under clause 39. However, the judge in that case did not accept that proposition and said:

> 'I am in no doubt at all that clause 63 can and should be construed without any suggestion that it is limited by clause 39 or that it should be preceded by a notice which is in some way identifiably referable to clause 39. The engineer and the employers have various options open to them under the contract and those options should not be restricted by the sort of argument that has been put in this case.
>
> I therefore find that the engineer has issued documents which on their face appear to put in motion the machinery of clause 63.'

Failing to proceed with due diligence

This is a matter which should be approached with the greatest of caution by engineers and employers. The courts have been most reluctant to impose on contractors any greater obligation than to finish on time.

In *Greater London Council* v. *Cleveland Bridge* (1986) the Court of Appeal refused to imply a term into a building contract that the contractor should proceed with due diligence notwithstanding the inclusion of that phrase in the determination clause of that contract. The point was repeatedly made in that case that the contractor should be free to programme his work as he thought fit. For cases showing the difficulty of defining due diligence and failure to proceed with it, see *Hill* v. *Camden* (1981) and *Hounslow* v.

Twickenham Garden Developments (1971).

Failure by the contractor to proceed in accordance with his approved clause 14 programme might provide some evidence to support a charge of failing to proceed with due diligence although failure to proceed to the programme is not itself a breach of contract.

Previous warnings

The reference to 'previous warnings' in clause 63(1) (b) (iv) should, perhaps, be read in conjunction with clause 46. That clause places a duty on the engineer to notify the contractor if he considers progress too slow to achieve completion by the due date.

Persistently or fundamentally in breach

The defaults of 'persistently or fundamentally in breach of his obligations under the Contract' may be mutually exclusive; that is, persistent breaches of minor obligations or a single breach which goes to the heart of the contract. It is not clear whether the 'previous warnings' apply to these defaults as well as to failing to proceed with due diligence.

Failure by the contractor to provide a bond as required by clause 10 could be an example of a fundamental breach; failure by the contractor to supply a programme under clause 14 would not be fundamental although persistent refusal to supply a programme might satisfy clause 63.

19.5 Notices under clause 63

Employer to give notice

For all defaults the employer must give seven days' 'notice in writing' specifying the defaults, before entering the site and expelling the contractor. The requirement to specify the default is new to the Sixth edition.

Another change from the Fifth edition is the provision for the employer to extend the 'period of notice' to give the contractor opportunity to remedy the default.

Even before this provision was added there were questions on whether, in the Fifth edition, the seven day period was a warning period for the contractor to remedy the default or a notice period for him to vacate the site. The wording of the Sixth edition leans heavily towards the first interpretation. If that is correct it suggests that the notice of determination is provisional upon the contractor not remedying the default. That in turn leads to difficult questions as to who is to judge whether the default has been remedied and what procedure is to be followed if there is a dispute.

To say that these are difficult issues is an understatement. In the case of *Attorney General of Hong Kong* v. *Ko Hon May* (1988), on conditions similar to the Fifth edition, both parties served notice of determination. The court held that the notices were provisional in the sense that their final validity would not be determined until arbitration, but pending the arbitration, providing they were given in good faith, they were both effective.

Engineer to certify in writing

For the performance defaults in clause 63(1)(b), before the employer can give notice of determination the engineer must certify in writing to the employer, with a copy to the contractor, that in his opinion the contractor is in default.

Again the question arises, can this be challenged and what happens in the meantime? Can the employer enter the site and expel the contractor or must he await the outcome of arbitration proceedings?

These were amongst the questions considered by the court in the *Tara Civil Engineering* case under an ICE Fifth edition contract. The engineer certified that Tara was in default and the employer gave notice of its intention to expel Tara from the site. Tara obtained an *ex parte* injunction restraining the employer from expelling them to which the employer responded by applying to the court for the order to be discharged. Granting the order requested by the employer, the court held that it should not go behind the engineer's certificate under clause 63 or any other documents relied on unless there was proof of bad faith or unreasonableness. The judge said:

> 'The most important of the three documents setting clause 63 in motion is the certificate of the engineer as to his opinion. It is important that the certificate is of his opinion only and not of

> fact. I take the view that I should only go behind that certificate, or behind any of the other documents relied on as setting in motion the clause 63 procedure, if there is either a lack of documents which on their face appear to set the procedure in motion or there is proof of bad faith or proof of unreasonableness.'

And later in the judgment, commenting on the policy of the courts, he said:

> 'At this stage there is no intention by the court to take sides in the determination of the ultimate disputes between the parties. The concern of the court is far from seeking to assist either party to break the contract. It is impossible to decide at this stage what is the conduct which would be in breach of the substantive terms of the contract. The court's present concern is to enforce the terms of the contract with regard to the only matters presently under consideration, namely, the regulation of the conduct of the parties pending the resolution of the substantive dispute by arbitration.'

Timing of the employer's notice

The Fifth edition was found to have a gap in procedure in that no time was specified for the employer to act after receiving the engineer's certificate.

In the case of *Mvita Construction* v. *Tanzania Harbours Authority* (1988), under FIDIC Conditions which matched those of the Fifth edition, the employer took three months before acting on an engineer's certificate. It was held that the employer was bound to give his notice within a reasonable time after the engineer's certificate to avoid a change in circumstances.

The Sixth edition now incorporates in clause 63(1) a provision that notice of determination is to be given as soon as is reasonably possible after receipt of the engineer's certificate.

19.6 Completing the works

Clause 63(2) which provides for the employer to enter the site and complete the works is a repeat of provisions in clause 63(1) of the Fifth edition.

The employer may either complete the works himself or employ another contractor. He can use for completion so much of the contractor's equipment and temporary works as is deemed to be the employer's property under clauses 53 and 54. And he can sell any of the contractor's equipment, temporary works, unused goods and materials and apply the proceeds towards the satisfaction of sums due to him from the contractor.

These are greater powers in theory than in practice. They would work to the full only if the contractor had title to all the equipment, temporary works and unused goods and materials. He may, in fact, have little title, with most equipment and temporary works on hire and most unused goods and materials not paid for.

The employer's claims may be valid against the contractor's legal successors but against third parties it has to be recognised that contractual provisions count for little.

19.7 *Assignment to the employer*

Clause 63(3), by which the engineer may require the contractor to assign to the employer the benefit of any agreement for the supply of goods, service or execution of work, is the same as the provision in the Fifth edition with the only change being that the notice period is altered from 14 to seven days.

Engineer and assignment

It is hard to understand why the engineer should be involved in the process of assigning agreements to the employer since it is very much a matter of commercial decision by the employer.

'The benefit of any agreement'

As to what is meant by assigning the benefit of any agreement, can it seriously be thought that any supplier or sub-contractor will provide goods or services to the employer with the burden of payment left with the contractor.

Almost certainly if any assignment is going to work in practice the employer will have to take on both the benefits and the burdens. And since the burdens will include payment of sums outstanding the employer will usually be better off simply by reaching new agreements with suppliers and sub-contractors.

19.8 Payment after determination

Clause 63(4) sets out a straightforward scheme where no payments are due from either party after determination until the expiration of the defects correction period or such later date as all the employer's expenses have been ascertained and the amount certified by the engineer.

These provisions, which match those in the Fifth edition, have been criticised as delaying the employer's right to recovery of sums due unnecessarily. In many cases the employer, by letting a new contract, will know well before completion what his loss is. With a common law determination his entitlement to recovery of loss would be immediate. Under these provisions the employer is caught with the contractor in what could be a prolonged wait for settlement.

Sums already certified

The employer can rightly refuse to pay on interim certificates coming due for payment after determination. Even if he has certificates which are overdue, as frequently happens when determination is anticipated and payments are held back, the employer should still be able to mount a successful counterclaim against any writ for payment.

Damages for delay

There is no certainty on the wording of clause 63(4) whether the damages for delay in completion are liquidated damages accrued to the date of determination plus general damages thereafter or liquidated damages for the full period to completion. It is suggested that the latter, which is more consistent with the provisions of clause 47 (liquidated damages), is probably correct.

Value of bonds

It is not clear from clause 63(4) how any sums the employer

obtains from security bonds are to be taken into account in settling the final sums due. The sums due are stated only in terms of amounts due on the contract, the costs of completion and expenses.

The engineer apparently has no power to consider payments made under bonds unless these are, perhaps, to be included as negative amounts within the scope of expenses.

19.9 Valuation at date of determination

The engineer is required by clause 63(5) to value the work at the date of determination by fixing:

(a) the amount earned by the contractor in respect of work done, and
(b) the value of unused goods and materials and any contractor's equipment and temporary works deemed to be the property of the employer under clauses 53 and 54.

This is not an easy task nor is it obvious what is its purpose. The amount which may eventually become due to the contractor is arrived at by the method of calculation given in clause 63(4) and the valuation at the date of determination is not included in this calculation.

The difficulties for the engineer in carrying out the valuation include deciding how to deal with temporary works and contractor's preliminaries and on what basis to value equipment.

19.10 Clause 64 – frustration

The purpose of clause 64 (frustration) is not to regulate a code of conduct or define frustrating circumstances. It is simply to ensure that the contractor is paid for the work he has done and is not forced by the Law Reform (Frustrated Contracts) Act 1943 into proving that the employer has received valuable benefit.

Frustration in construction contracts is, in fact, a most uncommon event, the test of which is a radical change of obligation.

In the case of *Davis Contractors* v. *Fareham UDC* (1956) Lord Radcliffe said:

> 'Frustration occurs whenever the law recognises that without default of either party a contractual obligation has become incapable of being performed because the circumstances in which performance is called for would render it a thing radically different from that which was undertaken by the contract. *Non haec in foedera veni.* It was not this that I promised to do.'

In that case a contract to build 78 houses in eight months took 22 months to complete due to labour shortages. The contractor claimed the contract had been frustrated and he was entitled to be reimbursed on a *quantum meruit* basis for the cost incurred. The House of Lords held the contract had not been frustrated but was merely more onerous than had been expected.

One of the few recorded cases of frustration being accepted in a construction contract in the UK is *Metropolitan Water Board* v. *Dick Kerr & Co. Ltd* (1918) where the onset of the First World War led to a two year interruption of progress. It was held that the event was beyond the contemplation of the parties at the time they made the contract and the contractor was entitled to treat the contract as at an end.

More recently in a Hong Kong case, *Wong Lai Ying* v. *Chinachem Investment Co. Ltd* (1979), a landslip which obliterated the site of building works was held by the Privy Council to be a frustrating event.

19.11 *Clause 65 – war clause*

The lengthy clause 65 providing for termination in the event of war has been carried virtually unchanged through the Fourth, Fifth and Sixth editions.

In short, the contractor is to continue to execute the works as far as possible for a period of 28 days and if the works are not then complete the employer is entitled to determine the contract.

By clause 65(3), clauses 66 and 68, settlement of disputes and serving of notices are preserved.

By clause 65(5), the contractor is given six headings of entitlement to payment:

(a) work executed at contract rates and prices
(b) preliminary items
(c) goods and materials ordered

(d) expenditure in anticipation of completion
(e) cost of repairing war damage etc.
(f) cost of removal of equipment.

Examination of clause 65 on a word by word basis will reveal enough inconsistencies to justify the clause heading but that has caused little trouble to anyone this last half century.

Chapter 20

Settlement of disputes

20.1 *Introduction*

This chapter examines clause 66 which provides an elaborate multi-stage process for the settlement of disputes. Reference is also made to two documents which are named in clause 66: the Institution of Civil Engineers Arbitration Procedure (1983) and the Institution of Civil Engineers Conciliation Procedure (1988).

In the Sixth edition, clause 66 has four components:

(a) a notice of dispute procedure
(b) a condition precedent to arbitration
(c) a semi-optional conciliation procedure
(d) an arbitration agreement and procedure.

The Fourth and Fifth editions each had a similar clause 66 dealing with the settlement of disputes but the notice of dispute and conciliation procedures are new to the Sixth edition. There have also been changes in the drafting of the provisions for the engineer's decision and for arbitration and these deserve close attention.

20.2 *'Except as otherwise provided'*

The purpose of the opening works of clause 66 'Except as otherwise provided in these Conditions' is not fully clear. These words were not included in the Fifth edition.

The words obviously apply to clause 70 on value added tax which expressly says that clause 66 shall not apply, and to clause 10 on performance bonds, but it is less certain whether there may be disputes under other provisions which are excluded. For example, if the contractor challenges the engineer for certifying under clause 63 that he is failing to proceed with due diligence, does that suspend the procedure of clause 63 until a clause 66 decision is given or does the procedure under clause 63 take precedence?

20.3 The engineer's decision

The common thread running through the various editions of the ICE Conditions is that the first stage of dispute resolution should be reference of any dispute to the engineer for a decision; and this, if not challenged within a specified time, should be binding on the parties.

This arrangement, although not unique, is unusual in that most standard forms of construction contract which have procedures for reviewing disputes prior to arbitration do so through an independent adjudicator or expert. It is certainly a fair question to ask what is the point of referring back to the engineer a dispute which in all probability arises from rulings he has previously made and which he is unlikely to reverse.

The standard answer, put forward in support of retention of the engineer's decision process, is that it works. No one has any figures by way of proof but it is said that there must be far more disputes than there are arbitrations in civil engineering so there must be a high level of acceptance of engineer's clause 66 decisions. A more jaundiced view might be that few contractors consider the cost of arbitration worth the risk and it is defeatism rather than acceptance which in reality ends most disputes.

Be that as it may, and however frustrating and time wasting the engineer's decision process may appear, it is in the contract and it must be meticulously observed by the parties and meticulously administered by the engineer.

Final and binding

Clause 66(4) which has the side note 'Effect on Contractor and Employer' states in reference to the engineer's decisions: 'Such decisions shall be final and binding upon the Contractor and the Employer unless and until....' The clause then goes on to describe the processes of conciliation and arbitration and the time limits within which action must be commenced.

This makes it essential that engineer's decisions are clearly recognisable as such and that referrals to the engineer are clear in their purpose and clear in what is in dispute.

In the case of *Monmouth County Council* v. *Costelloe & Kemple Ltd* (1965) the Court of Appeal had to consider whether a letter written by an engineer was a clause 66 decision under the ICE Fourth edition. If it was the contractor was time barred from commencing arbitration. In ruling that the letter was not a clause 66 decision because, amongst other things, the contractor had never asked for one, Lord Justice Harman said this:

> 'The other consideration which moves me is this. This is a process by which the defendants can be deprived of their general rights at law and therefore one must construe it with some strictness as having a forfeiting effect. It is not a penal clause, but it must be construed against the person putting it forward who is, after all, trying to shut out the ordinary citizen's right to go to the courts to have his grievances ventilated. Therefore, I think it would require very clear words and a very clear decision by the appointed person, namely the engineer, to shut the defendants out of their rights.'

As will be seen below under 'Notice of Dispute' the Sixth edition endeavours to avoid uncertain and unrequested clause 66 decisions. It remains good practice however that the parties and the engineer should endorse any correspondence under clause 66 with a specific reference to the clause.

The decision making process

The very fact that an engineer's clause 66 decision is different in status from his other decisions under the contract has led to the belief that the decision making process must be different.

There is nothing fixed on this. Some engineers do take representations from both parties on the matter in dispute; others do no more than review their files. The engineer is certainly not acting as an arbitrator, nor as a quasi-arbitrator – a misused phrase but explained by Lord Morris of Borth-y-Gest in *Sutcliffe* v. *Thackrah* (1974) as follows:

> 'There may be circumstances in which what is in effect an arbitration is not one which is in the provisions of the Arbitration Act. The expression quasi-arbitrator should only be used in that connection.'

The prime requirement in the decision making process is that the engineer must act fairly and impartially. It is essential therefore that if the parties do wish to make representations they are permitted to do so.

The prudent engineer will invite representations, not merely to cover his own position, but because by doing so he improves the prospects of the parties accepting his decision.

Directly employed engineers

Directly employed engineers may sometimes seem to be in a difficult position in reaching a balanced decision. This is how Mr Justice Macfarlan in *Perini* v. *Commonwealth of Australia* (1969) saw the matter (albeit under a different clause). He said:

> 'The second matter which I must mention is the entitlement of the Director to consider departmental policy. This point must be judged against a background that the Director is the senior officer of the department in New South Wales, that he is obliged to carry out the orders of his superiors and that he has many duties under this very agreement which he performs as the servant of the Commonwealth, and in the performance of which he is obliged to execute and give effect to departmental policy. I am of the opinion that in discharging the duties imposed upon him by clause 35 he is entitled to consider departmental policy but I am also of the opinion that he would be acting wrongly if he were to consider himself as controlled by it. His overriding duty in performing the function imposed by clause 35 is to give his own decision having regard to the rights and interests of the parties as I have described them. He is thus obliged to consider each application having regard to those rights and interests; he may also consider it from the point of view of departmental policy but the rights and interests must be the only matters involved in the decision. It is irrelevant if departmental policy coincides with the rights and interests of the parties under the agreement, but it would be quite wrong, in my opinion, for departmental policy to govern a particular decision, unless the personal decision of the Director having regard to the rights and interests of the parties under the agreement was that those rights and interests required it to be applied.'

The decision maker

The engineer is not permitted to delegate the making of his clause 66 decision and clause 2(4)(c) expressly prohibits this.

There can be, however, a fine line between taking advice from subordinates or colleagues and reaching one's own decision. The engineer who takes a hands-off role in the administration of the contract can be particularly vulnerable to the charge that he has delegated – and if such a charge can be upheld, the decision would, of course, be invalid.

For years many engineers and arbitrators have thought the practice of allowing the phrase 'This matter is being dealt with by Mr ...' to be used in letters on matters which could not be delegated was stepping too close to the line. However, in *Anglian Water Authority* v. *RDL Contracting* (1988), a case relating to a clause 66 decision under the ICE Fifth, Judge Fox-Andrews QC said this:

> 'In the commercial world many decisions are made by people such as Mr Rouse, who append their signatures to letters drafted by others. It would require compelling evidence to establish in such circumstances that the decision was not that of the signatory. The facts that Mr Baxter was the Project Engineer and had taken an active part previously in the contract had no probative value.'

The decision might well have been different had the engineer allowed a subordinate to sign in his name – a not unusual occurrence but one which should be avoided.

Taking advice

On the matter of the engineer's taking advice this further extract from the judgment in the *Perini* case is worth noting, Mr Justice Macfarlan said:

> 'I cannot accept all the arguments submitted by learned counsel for the plaintiff that the Director is bound to investigate every dependent fact himself; this conclusion would, I think, be to ignore the realities of the situation. I am of opinion, though, that by this agreement and by his mandate he may act upon the findings and opinions of other persons, be they subordinates or

independent persons such as architects or meteorological observers; he may also consider and pay attention to the recommendations of subordinates with respect to the very application he is considering. I do agree though that the actual decision must be one which flows from the volition of his own mind and I am of the opinion that it is quite irrelevant that that decision is expressed by the placing of his initials upon the recommendation of a subordinate officer.'

20.4 Disputes

Types of disputes

What is meant by 'a dispute' and when does it arise?

Clause 66(2) of the Sixth edition says that a dispute shall be deemed to arise when one party serves on the engineer a notice in writing stating the nature of the dispute. Rule 1 of the ICE Arbitration Procedure (1983) states that a dispute or difference shall be deemed to arise when a claim or assertion made by one party is rejected by the other party and that rejection is not accepted.

These are not necessarily the same thing. The first requires communication between a party and the engineer; the second requires communication between the parties. The first may be described as a dispute which can be referred to the engineer for his decision; the second, as a dispute which can be referred to arbitration.

Support for the proposition that there are two kinds of dispute comes in clause 66(1) which refers to any dispute between the employer and the contractor and any dispute as to any decision, certificate etc. of the engineer. The problem with this is that the drafting of clauses 66(5) and 66(6) implies that any dispute which has been decided by the engineer is a dispute between the parties which is capable of being referred to conciliation or arbitration.

In practice if the engineer has decided on a matter referred to him unilaterally – as is common practice under the Fifth edition where similar difficulties exist – then all the parties have is a decision. It remains to be discovered if they have, or ever have had, a dispute.

There is certainly much to be said for the parties exchanging their views on an issue before either serves a formal notice of dispute under clause 66.

Scope of dispute

Arbitration clauses in some contracts have been held to be limited in their scope with the courts distinguishing between disputes arising 'under' a contract from disputes arising 'in connection' with a contract. The wording of clause 66(1) is wide enough to avoid most argument in its reference to:

(a) a dispute of any kind between the employer or the contractor in connection with or arising out of the contract or the carrying out of the works
(b) including any dispute as to any decision, opinion, instruction, direction, certificate or valuation of the engineer
(c) whether during the progress of the works or after their completion and whether before or after determination, abandonment or breach of contract.

It is suggested that the words which commence 'whether during' which are placed in brackets in clause 66(1) are intended only to apply to engineer's decisions as confirmation that he is not *functus officio* after completion or determination. If the words have wider application it might be held that disputes arising out of matters which occur before the contract are not within the scope of clause 66.

Misrepresentation

There has, of course, long been a view that such matters do not fall within the scope of an arbitration clause whatever its wording. As recently as 1987 in *Blue Circle Industries* v. *Holland Dredging Co.* (1987), a case on a contract based on the ICE fifth edition and with the same clause 66, Lord Justice Purchas said this:

> 'I would have held, had it been necessary to do so, that the claims for damages for negligence (ground (a)) and misrepresentation (ground (d)) did not arise 'out of the contract or the carrying out of the works' (clause 66). The only question which

> needs further consideration is whether these claims arose 'in connection with ... the contract' (clause 66) or 'out of or in connection with this order' (clause 13). Counsel for the respondents submitted that all the issues and allegations comprised in the negligence and misrepresentation claims were repeated in the claims for breach of contract and negligent execution of the contract. With respect to this attractive argument I consider that it has a fallacy which can be shortly stated. If it were not for the allegations of negligent advice or misrepresentation, if these are substantiated, there would never have been a contract and, therefore, there would not have been an agreement to arbitrate. To refer these matters to an arbitrator would in effect be inviting him to adjudicate upon his own jurisdiction and this is not in accordance with the authorities.'

However, within months, those observations, which were given *obiter* (comments in passing and not binding as precedent), were considered as 'not persuasive' by Lord Justice Bingham in *Ashville Investments* v. *Elmer Contractors* (1987). In that case it was held that claims arising out of alleged innocent or negligent misstatements were claims arising 'in connection with' the contract.

Notice of dispute

Clause 66(2) which formalises the procedure for giving notice of dispute is new. It should prevent inadvertent use of clause 66.

The 'Notice of Dispute' is to be in writing; it is to state the nature of the dispute; and it is to be served on the engineer. Curiously there is nothing in clause 66(2) to oblige the party serving notice to serve notice on the other party as well as on the engineer. This suggests either that a notice of dispute should not be served until the parties have established that they are in dispute or that the engineer has a duty to consult both parties in reaching his decision. It would certainly be an abuse of the procedure if the first one party knew of the dispute was upon receipt of the engineer's decision.

Other steps or procedures

The provision in clause 66(2) that no 'Notice of Dispute' may be served until other steps or procedures in the contract have first been taken may be another case of good intentions creating complications. A straightforward example would seem to be that a dispute on an engineer's representative's act or instruction must go through clause 2(7) before clause 66. But does that necessarily apply to acts or instructions of the engineer's representative made using delegated powers of the engineer?

More problematical is, when and by whom can challenges be made to non-observance of the procedure? Is it possible for the engineer to reject a notice of dispute if he considers that other steps and procedures have still to be observed? Can one of the parties challenge an engineer's decision as invalid for non-observance of the preliminary steps? Could the authority of an arbitrator be challenged on similar grounds?

If an engineer's decision can be rendered invalid by non-observance of prior steps or other procedures, and lawyers may well argue that proposition if they find themselves time-barred elsewhere, then the point made earlier in this chapter that no 'Notice of Dispute' should be served until the parties have established a dispute between themselves is greatly strengthened.

20.5 Time limits for decisions

Clause 66(3) requires the engineer to state his decisions in writing and to give notice of the same to both employer and contractor within the time limits set out in clause 66(6).

In fact, clause 66(6) does not directly place time limits on the engineer. It says if the engineer fails to give a decision within one calendar month after service of a notice of dispute, where the certificate of completion for the whole of the works has not been issued, or within three calendar months where the certificate has been issued, then either party may refer the dispute to arbitration.

This matter came up in the case of *ECC Quarries Ltd* v. *Merriman Ltd* under an ICE Fifth edition contract. The engineer failed to give his clause 66 decision within three months of being requested to do so by the contractor. The contractor then wrote to the engineer giving him a reasonable time to deal with the matter. The engineer's decision, which should have been given in May, was eventually given in July. Nine months later the contractor sought to refer the matter to arbitration. The employer successfully obtained from the court a declaration that the contractor was precluded from pursuing his claim as he had failed to give notice to refer within three months of the decision.

Some interesting points came out of the case. The judge held that, if the engineer had purported to give his decision after either party had referred the matter to arbitration, the decision would be null and void because the engineer would have been rendered *functus officio* at the time of the referral. Late delivery of the engineer's decision did not itself however render the decision null and void. The time for giving a decision could be effectively extended by one party acting unilaterally – but note that in this case there was no opposition on that particular point.

Engineers would be well advised to seek the agreement of both parties if they need an extension of time for their decision and the parties also would also do well to clarify their respective positions.

In the *ECC* case the employer wisely had the position clarified by the court rather than risk a costly arbitration which would be vulnerable to appeal.

20.6 Effect of engineer's decision

The most significant provision in clause 66(4) is that a decision of the engineer is final and binding upon the parties unless and until either:

(a) the recommendation of a conciliator is accepted, or
(b) the decision is reversed by an arbitrator.

This has the effect that if the parties fail to operate the conciliation or arbitration procedures in time they are left with an engineer's decision which is permanently binding.

The other provisions of clause 66(4) are:

(a) that the contractor shall continue to proceed with the works with all due diligence, and
(b) that the contractor and the employer shall both give effect forthwith to every decision of the engineer.

The intention of the first of these is perhaps to do no more than emphasise that the contractor is not entitled to down tools pending an engineer's decision. The second prevents the parties ignoring an engineer's decision pending conciliation or arbitration. Thus if it is a payment certificate which is in dispute and the engineer has decided in the contractor's favour against a challenge from the employer, then the employer must pay forthwith.

In practical terms the contractor might have some difficulty in enforcing payment even with the support of an engineer's decision. The ICE Conditions give the contractor no express rights of determination other than for prolonged suspension under clause 40 and it is doubtful if the courts would entertain an application for summary judgment under Order 14 of the Rules of the Supreme Court if the employer applied for a stay of proceedings under Section 4 of the Arbitration Act 1950.

In *Enco Civil Engineering Ltd* v. *Zeus International Development Ltd* (1991) the employer challenged an engineer's payment certificate and declined to pay, and asked for a decision under clause 66. Before the decision was given the contractor issued a writ for summary judgment on the amount certified. The employer issued a summons to stay proceedings. It was held that the procedure in clause 66 did not prevent the court from ordering a stay; that summary judgment could not be given when the certificate was under review; and that the employer had arguable defences and there was nothing in the ICE Conditions to prevent him setting off cross-claims against sums certified.

The difficulty of applying the provisions of clause 66(4) to a dispute under clause 63 (determination) has been considered in chapter 19.

20.7 Timescale for conciliation and arbitration

Arbitration

Clause 66(6) provides the timescale for reference to arbitration.

Such reference must be made within three calendar months of receipt of the engineer's decision or the last date when such decision should have been given; or within one calendar month of the receipt of a conciliator's recommendation.

The courts do have a discretionary power under section 27 of the Arbitration Act 1950 to extend time for commencing arbitration proceedings. In *McLaughlin & Harvey plc* v. *P & O Developments* (1991) a contractor who gave notice to refer to arbitration a day late was granted relief because there was no possible prejudice to the employer and the hardship that the contractor would suffer would be out of proportion to his fault.

In the case where the engineer fails to give his decision on time, perhaps after indicating that he is unable to do so, the wording of clause 66(6) is possibly restrictive in requiring notice to refer to be given within three months of the due dates notwithstanding any relaxation which may have been granted to the engineer. This is a point which the parties need to clarify in granting any relaxation.

Conciliation

Clause 66(5) does not place any direct time limits on reference to conciliation. The only certain barrier is when a notice to refer to arbitration has already been served.

There can be little doubt that the draftsmen intended the same time limits to apply to conciliation as to arbitration. However, the wording of clause 66(5) does not properly achieve this, although it has been suggested that it can be implied.

The problem created by the deficient wording is that if neither party gives notice to refer to arbitration then either may be free, within the legal limitation periods of six or twelve years, to refer a dispute to conciliation and then within one month of the recommendation to refer to arbitration – thereby completely defeating the time limits in clause 66(6).

The other party would seem to be powerless to prevent this. Normally it will be the party who has applied for the engineer's decision who will be dissatisfied and who will refer to conciliation or arbitration. The other party remains passive until such notice is served. The opportunities in this for tactical delay are obvious.

Clause 66(5) needs to be amended to close the loophole.

20.8 Conciliation

The introduction of conciliation into the Sixth edition as a step in the procedure for resolving disputes is courageous and forward looking. Alternative dispute resolution (ADR) in its various forms is gaining ground as dissatisfaction grows with the costs of litigation and arbitration but it is still in the early stages of development in the UK and there is little familiarity with terminology or procedures.

What is conciliation?

It may come as a shock to civil engineers, well versed in precise definitions, to learn that the language of ADR is fluid and there are no fixed meanings to the terms conciliation and mediation. What half the world and half this country calls conciliation the other half calls mediation.

Attempts are being made to fix definitions in some quarters but against this there is the argument that flexibility is the vital ingredient of ADR and the various processes should be left to converge and adapt to suit individual circumstances.

Sir Lawrence Street, former Chief Justice of New South Wales in a paper to the 1991 Annual Conference of the Chartered Institute of Arbitrators said this:

> 'Attempted distinctions between conciliation and mediation, and between active and passive, if they exist at all, could at times become extremely fine. I myself hold firmly to the view that there are no such relevant, practicable or useful distinctions: the words mediation and conciliation are, I suggest, synonymous.'

The Master of the Rolls, Lord Donaldson, in an address in 1991 to the London Common Law and Commercial Bar Association, did offer these definitions:

> 'Conciliation: This is a process whereby a neutral third party listens to the complaints of the disputants and seeks to narrow the field of controversy .... He moves backwards and forwards between the parties explaining the point of view of each to the other. He indulges in an onion peeling operation. He peels off each individual aspect of complaint, inquiring whether that aspect really matters. In the end he and the parties are left with a core dispute which, so much having been discarded under the guidance of the conciliator, at once seems more capable of settlement on common sense lines.
>
> Mediation: This what a PR man would describe as 'conciliation plus'. The mediator performs the functions of a conciliator, but also expresses his view on what would constitute a sensible settlement. In putting forward his suggestion, which the parties will be free to reject, the mediator will in most cases be guided by what he believes would be the likely outcome if no settlement was reached and the matter went to a judicial or arbitral hearing.'

Lord Donaldson went on in that address to make the valuable but little expressed point that neither conciliation, mediation nor other like process are 'alternative' forms of dispute resolution. They are additional forms of dispute resolution in that they are only part of a process which in the absence of agreement will lead to an imposed decision.

ICE Conciliation Procedure 1988

Clause 66(5) of the Sixth edition permits either party, if no notice to refer to arbitration has been served, to require that the dispute be considered under the ICE Conciliation Procedure 1988 or any amendment or modification thereof.

The recommendation of the conciliator is deemed to be accepted unless notice to refer to arbitration is served within one calendar month of its receipt.

Under the rules of the Conciliation Procedure, the conciliator is to use his best endeavours to conclude the conciliation as soon as possible and, in any event, within two months of his appointment. The rules are to be applied in a manner most conducive to efficient conduct with the primary aim of obtaining the conciliator's recommendations as soon as possible.

The parties may send to the conciliator and each other written submissions of their case and the conciliator may generally inform himself in any way he thinks fit of the nature and facts of the dispute. He may convene a meeting and take evidence, but is not bound by rules of evidence or procedure. He may at any time if it is appropriate or if he is requested by the parties express his preliminary views on the matter.

Within 21 days of any meeting the conciliator shall prepare his recommendations as to the way the matter should be settled and, if he considers it appropriate, he may state his opinion of the matter and his reasons thereon.

What is the ICE Procedure?

It has to be said that the ICE Procedure is not obviously either conciliation or mediation as defined by Lord Donaldson; it outlines a more formal process. However, training for the ICE list of conciliators does encompass both the conciliation and mediation techniques described by Lord Donaldson so it is inevitable that for some time conciliation under the ICE

Conditions will be a flexible and possibly uncertain process.

It is understood that the Institution of Civil Engineers is aware of the need to clarify the process, not least so that those participating will not be taken by surprise at finding emphasis placed in practice on negotiation rather than evidence taking.

A new conciliation procedure, which may even be named a mediation procedure, will probably be issued in 1993.

Will conciliation work?

Conciliation is intended to be short, sharp and cheap. The essence is consensual. It may not achieve the settlement the parties want but it is said to have succeeded if it achieves a settlement the parties can live with.

There are two aspects of the incorporation of conciliation into the ICE Conditions which cause some concern.

Firstly, it is not truly optional. Indeed it will often be compulsory for the non-claiming party since on receipt of an engineer's decision that party is not likely to call for either conciliation or arbitration. The option, therefore, on whether to conciliate or arbitrate is effectively only in the hands of the claiming party. The question then is, will conciliation work when one party is there by compulsion?

Secondly, there is the practical point that once an understanding of what conciliation is all about develops in the industry contractors will see that it is a no-loss procedure. That is to say, they can make a claim; call for an engineer's decision; and then require the employer to go through a negotiating/conciliation process. This is not the ideal recipe for reducing claims and discouraging disputes.

20.9 Arbitration

Whatever happens in conciliation it will only lead to a binding settlement if the parties so agree. They are free to reject any recommendation and pursue their legal remedies.

In the ICE Conditions, the legal remedy is fixed by clause 66 as arbitration. Arbitration is consensual – one party cannot force another to settle a dispute by arbitration unless the contract contains an arbitration agreement covering the subject matter, or both parties make a separate such agreement subsequently.

The Arbitration Acts 1950, 1975 and 1979 provide a statutory basis on which difficulties can be resolved, e.g. powers of the arbitrator, refusal of the arbitrator to act, appeal from an arbitrator's award and much else. The arbitrator's powers are derived not only from these statutes and supporting case law, but from the terms of the arbitration agreement itself.

If there is a valid arbitration agreement covering the dispute then the courts will normally stay any legal proceedings brought by one party on the motion of the other, thus upholding the consensual agreement – section 4 Arbitration Act 1950. The courts will not grant a stay however if the applicant has taken a step in the legal action, that is, shown himself willing to contest the legal proceedings.

Arbitration v. Litigation: Advantages and Disadvantages

The principle differences can be summarised by listing the essential characteristics of arbitration:

(a) the hearing is private
(b) the award is published privately
(c) the parties may be able to choose their own arbitrator
(d) the venue of the hearing is flexible
(e) the hours, date and all administrative details of the hearing are flexible
(f) the hearing can be less formal
(g) the rights of appeal against an award are restricted
(h) not ideal for multi-party disputes
(i) the arbitrator and the venue have to be paid for
(j) lawyers are optional
(k) lay advocates are optional
(l) the arbitrator has no general power to order security for costs but may have under some rules
(m) a reluctant party may be less robustly dealt with
(n) can be cheaper and quicker
(o) documents-only may be an alternative option
(p) the arbitrator has power to award interest and costs
(q) the arbitrator can be asked to include site visit etc.
(r) the arbitrator may be given powers the courts do not have
(s) arbitration may be imposed if agreement included the contract
(t) the legal aid scheme is not applicable.

Power to open-up and review certificates

Clause 66(8)(a) of the Sixth edition gives the arbitrator full power to open up, review and revise any decision, direction, certificate or valuation of the engineer.

The decision in *Northern Regional Health Authority* v. *Derek Crouch Construction Co. Ltd* (1984) means that unless an engineer has wilfully refused to operate the contract machinery, then the arbitrator alone (having such powers bestowed on him by the arbitration agreement in the contract) can open up and review decisions, awards and certificates. That is to say, the courts will not deal with such matters. However, the effect of the *Crouch* decision has now been diminished by the Courts and Legal Services Act 1990 which enables the parties, by agreement, to confer the powers of the arbitrator on the courts.

ICE Arbitration Procedure 1983

Clause 66(8)(a) requires that any reference to arbitration shall be conducted in accordance with the ICE procedure in force at the time of the appointment of the arbitrator.

The 1983 Procedure was drafted to overcome the problem that arbitration with legal representatives on both sides was becoming indistinguishable from litigation and it aims to bring out and develop those features of arbitration which distinguish it from and give it an advantage over litigation. It does so by creating a new range of powers available to the arbitrator and a new set of procedures for particular types of disputes.

Powers of the arbitrator

The express powers conferred by the ICE Procedure include:

Rule 5 – power to control the proceedings
Rule 6 – power to order protective measures and deposits of money for security
Rule 7 – power to order concurrent hearings
Rule 8 – powers at the hearing
Rule 9 – powers to appoint assessors and to seek outside advice
Rule 14 – powers to make summary awards
Rule 16 – power to put questions and order tests.

The new procedures are:

Part F – Short Procedure
Part G – Special Procedure for Experts
Part H – Interim Arbitration.

Short Procedure

The short procedure is appropriate for cases when there is a well defined dispute. Each party sets out his case in a file for submission to the other party and the arbitrator, and within a month the parties appear before the arbitrator to make oral submissions and answer the arbitrator's questions. There is no cross-examination of witnesses and each party meets his own costs.

Special Procedure

The special procedure for experts is based on the short procedure but is designed to apply to limited issues which depend upon expert evidence. After receipt of each party's file and after viewing the site if he thinks it necessary, the arbitrator meets the expert witnesses who may address him and question each other. Legal representation is not permitted.

Interim Procedure

The interim procedure applies when the arbitrator is appointed and the arbitration proceeds before completion of the works (except in the case of a dispute arising under clause 63). The procedure lays down a tight timescale and allows the arbitrator to make interim awards, findings of facts, summary awards and interim decisions.

Standard notices

The Institution of Civil Engineers has produced a set of standard notices for use with the ICE Arbitration Procedure.

The notices deal with:

(a) contract details

(b) references to arbitration
(c) appointment of the arbitrator by the parties
(d) appointment of the arbitrator by the President of the Institution of Civil Engineers.

20.10 Other aspects of clause 66

Appointment of arbitrator

No time limit is fixed in the ICE Conditions or in the ICE Arbitration Procedure for appointment of the arbitrator after notice to refer has been served.

This can lead to lengthy dormant periods while the parties take stock of their positions but it is open to either party to make the first move. This is not the prerogative of the claiming party.

Clause 66(6) (a) uses the phrase 'to the arbitration of a person to be agreed upon by the parties' in relation to time limits but then clause 66(7) goes on to set out the procedures when the parties fail to agree.

Under clause 66(7) (a) if the parties have failed to appoint within one calendar month of a notice to concur in the appointment of an arbitrator, either party can apply to the President of the Institution of Civil Engineers to make an appointment.

Evidence in arbitration

Under clause 66(8) (b) neither party is limited in arbitration proceedings to the evidence and arguments put before the engineer for the purpose of obtaining his clause 66 decision.

This often comes as a matter of surprise to one or both of the parties and not infrequently there is talk of sharp practice. But although there is nothing to stop the parties holding back their case before the engineer, or for that matter, before the conciliator, the common reason for new evidence and arguments is the involvement of lawyers and claims consultants who, quite rightly, want to do their best for their client unencumbered by what has gone before.

Binding on the parties

Under clause 66(8) (c) it is stated, more as a formality than a contractual provision, that the award of the arbitrator shall be binding on all the parties.

This would not be effective in preventing an appeal on a point of law or an application to the courts to have the arbitrator's award set aside.

Arbitration before completion

The Sixth edition permits, as did the later version of the Fifth edition, arbitration to proceed notwithstanding that the works are not complete – see clause 66(8)(d).

The words in the clause 'unless the parties otherwise agree' cover the situation where notice to refer has been given to avoid a time-bar but the claiming party is in no urgency to pursue the matter.

Engineer as witness

Clause 66(9) confirms that the engineer may be called as a witness to give evidence in the arbitration notwithstanding his involvement in giving his clause 66 decision.

The engineer would normally only be called as a witness of fact on a dispute in which he was involved and it would be unusual for him to appear as an expert witness.

20.11 Summary

As long as clause 66 remains in the ICE Conditions there will be debate on whether it is over-elaborate and artificial in its premise that the engineer is the right person to adjudicate on disputes. It is certainly inconceivable that any new form of construction contract would adopt such a procedure. Perhaps as the ICE Conditions face increased competition over the next few years from other forms of contract it will be possible to draw comparisons on this and other contentious provisions, and users of the Conditions will be able to form their own opinion on whether the 'thorough review' which led to the Sixth edition went far enough in this and other matters.

Appendix

ICE Conditions of Contract – Sixth edition: Improvement analysis

* indicates significant improvement or change in the balance of risk

Clause no.	Change from Fifth edition	Improvement to contractor	Improvement to employer
1(1) (l)	provisional sum applies only to specific contingency	x	
1(1) (v)	other places may be designated as forming part of the site	x*	
1(5)	cost means all expenditure properly incurred		x
1(5)	cost includes finance charges	x*	
1(5)	cost does not include profit		x
2(1) (a)	engineer to carry out the duties specified or implied	x*	
2(1) (b)	particulars of engineer's authority requiring consent		
2(1) (c)	engineer no authority to amend terms and conditions		x
2(2) (a)	engineer to be a named individual	x	
2(2) (a)	engineer to be a chartered engineer		

Clause no.	Change from Fifth edition	Improvement to contractor	Improvement to employer
2(2)(b)	replacement of named chartered engineer		
2(4)	general delegation of powers no longer included	x	
2(5)(a)	assistants can carry out delegated duties of the engineer's representative		
2(5)(b)	instructions of assistants to be in writing	x	
2(6)(a)	instructions of engineer or engineer's representative to be in writing	x	
2(6)(b)	oral instructions if confirmed by contractor are deemed to be written instructions	x	
2(6)(c)	engineer or engineer's representative to specify in writing under which clause any instruction is given	x	
2(8)	engineer to act impartially	x	
3	employer not to assign without consent	x	
3	consent to assignment not to be unreasonably withheld	x	
4(1)	contractor can sub-contract whole of the works with consent of the employer	x	
4(2)	contractor may sub-contract parts of the works without approval	x*	
4(5)	engineer has power to order removal of sub-contractors		x
6(1)	contractor entitled to four copies of contract documents		

Clause no.	Change from Fifth edition	Improvement to contractor	Improvement to employer
6(2)	contractor to supply engineer with four copies of contractor design documents		
6(2)	documents need not be returned on completion		
6(3)	engineer and employer have free power to reproduce contractor's design documents		
7(2)	contractor to provide further documents as required by engineer		
7(4)(b)	delay in issue caused by contractor's default		
7(6)	permanent works designed by contractor		
7(7)	engineer's approval does not relieve contractor of his responsibilities		x
7(7)	engineer responsible for integration and co-ordination of contractor's design	x	
8(2)	contractor to exercise reasonable skill and care in his design	x	
9	execution of contract agreement expressed optionally		
10	alternative to bond of two sureties omitted		
11(1)	employer to make available all information on ground conditions	x*	

Clause no.	Change from Fifth edition	Improvement to contractor	Improvement to employer
11(2)	inspection and examination only to be so far as practicable in connection with extent and nature of work, materials, communication, access	x*	
11(3)	contractor deemed to have based his tender on information made available by employer	x*	
12(1)	contractor to give notice of encountering unforeseen conditions		x
12(4)	engineer can require contractor to investigate and report on alternative measures		
12(6)	contractor can claim profit on all extra costs	x	
13(3)	contractor entitled to profit on additional work	x	
14(1)(a)	contractor's programme to show order having regard to provisions of clause 42(1)		x
14(1)(b)	contractor to submit general description and methods of construction with programme		x
14(1)(c)	contractor to submit a revised programme following rejection		x
14(2)	procedure for accepting or rejecting contractor's programme		
14(3)	engineer to respond to further information within 21 days	x	

Clause no.	Change from Fifth edition	Improvement to contractor	Improvement to employer
14(4)	procedure for revised programme		
14(7)	engineer to state within 21 days whether or not he consents to the contractor's proposed methods	x	
14(8)	contractor entitled to profit on additional work	x	
15(1)	contractor to superintend during construction and completion		
16	contractor's employees' removal for lack of regard to safety requirements in the contract		
19(1)	contractor to comply with engineer's representative's requirements on safety		
20(1)	contractor's responsibility for care of works includes materials, plant and equipment		
20(1)	contractor's responsibility commences on works commencement date		
20(1)	contractor's responsibility ends on date of issue of certificate of substantial completion	x	
20(2)	excepted risks defined as loss or damage to the extent it is due to specified events		x
20(3)	reworded provisions for rectification of loss or damage		x
21(1)	insurance of the works to be full replacement cost plus 10%		x

Clause no.	Change from Fifth edition	Improvement to contractor	Improvement to employer
21(1)	contractor not required to insure his construction plant	x	
21(2) (d)	amounts not insured, including excess, recoverable according to responsibility	x	x
22(2) (c)	omission of engineer from exceptions to contractor's liability		
23(1)	duration of insurance cover not specified		x
23(1)	insurance to be in joint names	x	x
23(2)	insurance policy to include a cross liability clause	x	x
25	requirement for insurance to be with an approved insurer dropped	x	
26(3) (b)	'contract' substituted for 'drawings, specifications'		
26(3) (c)	specified temporary works excluded from employer's warranty		
28(1)	the contractor's indemnity on patent rights amended so that employer now indemnifies the contractor in respect of infringements resulting from the design or specification	x	
29(2)	'stated' replaces 'specified'		
29(3)	contractor's indemnity for noise, disturbance or other pollution excludes unavoidable consequence of constructing the works	x	

Clause no.	Change from Fifth edition	Improvement to contractor	Improvement to employer
29(4)	employer indemnifies contractor against claims which are the unavoidable consequence of carrying out the works	x*	
31(1)	engineer's representative can issue requirements on reasonable facilities		
31(2)	contractor entitled to profit on additional work	x	
34	not used		
35	engineer's representative can accept returns of labour and contractor's equipment		
39(1)(c)(ii)	engineer's power to order removal of work designed by the contractor not in accordance with the contract		x
40(1)	contractor entitled to profit on additional work	x	
41(1)	works commencement date to be specified, notified within 29 days, or agreed	x	
42(1)	contract may prescribe nature of access to be provided by the employer and may prescribe order of construction		x
42(2) (a)	employer to give access as specified	x	
42(2) (b)	employer to give possession in accordance with accepted clause 14 programme		x
42(3) (b)	contractor entitled to profit on additional work	x	
42(3)	engineer to notify contractor and copy to employer		

Clause no.	Change from Fifth edition	Improvement to contractor	Improvement to employer
42(4)	'access' replaces special or temporary wayleaves in connection with access		
42(4)	additional 'facilities' replaces additional 'accommodation' outside the site		
43	works to be substantially completed within the time specified		
44(1)	clause reworded so that delay is applicable to all entitlements to an extension		
44(2)	new procedure for assessment of delay	x*	
44(3)	interim extensions of time to be granted forthwith	x*	
44(4)	assessment at due date for completion to be within 14 days	x	
44(5)	final determination of extension to be within 14 days	x	
46(1)	phrase 'Clause 43 and 44' added		
46(2)	permission to work nights or Sundays applies only to site		
46(3)	new provision for employer to request accelerated completion	x	x
47(1)	provision for liquidated damages to be lesser sum than genuine pre-estimate omitted		

Clause no.	Change from Fifth edition	Improvement to contractor	Improvement to employer
47(1)	added reference to acceleration provision in clause 46(3)		
47(2)	reworded provisions for liquidated damages for sections		
47(4) (a)	provision for limitation of amount of liquidated damages	x	
47(4)	engineer's notification no longer a condition precedent to the deduction of damages		x
47(4) (b)	if sum omitted from appendix or stated as 'Nil', damages not payable	x	
47(6)	new provision for suspension of damages when contractor is in culpable delay	x*	
48(1)	contractor's notice he considers works to be substantially completed no longer stated to be a deemed application for a certificate		
48(1)	outstanding work to be finished in accordance with provisions of clause 49(1)		
48(2)	contractor entitled to a certificate of substantial completion where premature use by employer	x	
48(4)	'considers' replaces 'of the opinion' in relation to substantial completion of parts of the works		

Clause no.	Change from Fifth edition	Improvement to contractor	Improvement to employer
49(1)	undertaking on outstanding work can specify a time for completion		x
49(2)	reworded obligation to execute work of repair 'within 14 days'		
51(1)(a)	'is in his opinion necessary' replaces 'may in his opinion be necessary'		
51(1)(b)	'may order any variation' replaces 'shall have the power to order'		
51(1)(b)	'desirable for the completion and/or improved functioning' replaces 'desirable for the satisfactory completion and functioning'		
51(1)	'any specified sequence' replaces 'the specified sequence'		
51(1)	'required by the contract' replaces '(if any)'		x
51(1)	variations may be ordered during the defects correction period		x
51(2)	reworded provision on oral instructions	x	
51(3)	variations necessitated by contractor's default not to be taken into account in ascertaining contract price		x
52(1)(b)	valuation of variations ordered during the defects correction period	x	
52(3)	engineer's power to order daywork		

Clause no.	Change from Fifth edition	Improvement to contractor	Improvement to employer
52(4) (b)	reference made to clause 56(2)		
52(4) (b)	claims to be notified within 28 days of the happening of the event		x
53(1)	greatly shortened provision for vesting of contractor's equipment	x	
54(1)	engineer can direct transfer of goods and materials off-site to employer		x
56(4)	procedure for payment for dayworks simplified	x	
57	'Unless otherwise provided in the Contract' replaces 'Except where any statement'		
57	'or any other statement' added as proviso to Standard Method of Measurement		
58	clause re-written with definitions excluded		
58(2) (b)	'himself' added after contractor		
59(1) (d)	new ground for objection to nominated sub-contractor in respect of security for performance	x	
59(2)	now covers engineer's action on both objection and termination		
59(2)	engineer's power to direct contractor to enter into a nominated sub-contract omitted	x	

Clause no.	Change from Fifth edition	Improvement to contractor	Improvement to employer
59(2)	engineer can instruct contractor to secure a sub-contractor of his own choice	x	
59(3)	the exception for nominated sub-contractor's default is omitted		x
59(4)	the employer's indemnity against the nominated sub-contractor's default is omitted		x
59(4) (b)	if the engineer refuses consent to terminate the contractor is entitled to instructions under clause 13	x	
59(4) (c)	engineer's action on termination reworded		
59(4)	termination without the employer's consent not covered		
59(4) (d)	reworded provision for steps to be taken to recover from nominated sub-contractor following termination		
59(4) (e)	reworded provision on reimbursement of contractor's loss		
60(1)	contractor to submit statements at monthly 'intervals'	x	
60(3)	payments due at monthly intervals after the certificate of completion of the whole of the works and not subject to minimum limits	x	
60(5)	clause reworded to refer to retention limits in the appendix		

Clause no.	Change from Fifth edition	Improvement to contractor	Improvement to employer
60(6)	clause reworded to ensure the employer has half the retention at substantial completion for the whole of the works		x
60(7)	interest on overdue payments due if arbitrator holds sums should have been certified by a particular date	x*	
60(9)	engineer to supply detailed explanation of certificate		
60(10)	employer to notify contractor where payments differ from amounts certified	x	
61(1)	phrase 'and maintain' omitted in respect of issue of certificate		
62	urgent repairs can be ordered by engineer's representative		x
63(1)(a)(i)	administration order listed as additional default		x
63(1)(a)(ii)	28 day period of grace on execution levied on goods	x	
63(1)	sub-contracting without consent omitted as a default	x	
63(1)	employer must specify defaults in termination notice	x	
63(1)	employer may extend period of notice	x	
63(1)	notice of determination to be given as soon as reasonably possible after engineer's certificate	x	

Clause no.	Change from Fifth edition	Improvement to contractor	Improvement to employer
63(3)	period for assignment altered from 14 to 7 days		
66(1)	new opening words 'Except as otherwise provided in these Conditions'		
66(2)	new notice of dispute procedure	x	x
66(4)	effect of engineer's decision altered to allow for conciliation		
66(5)	new conciliation procedure	x*	
66(6)	provisions for arbitration amended to take account of conciliation		
67(2)	application for clause 66 to Scotland reworded		
68	procedure for notices omits references to sending by post or leaving at addresses		
69	provisions for labour-tax fluctuations reworded		
70	provisions for value added tax reworded and simplified		
71	clause on metrication omitted		

Table of Cases

Note: references are to chapter sections.
The following abbreviations of law reports are used:

AC	Law Reports Appeal Cases Series
All ER	All England Law Reports
BIN	Building Industry News
Bing	Bingham's Reports, Common Pleas
BLM	Building Law Monthly
BLR	Building Law Reports
Camp	Campbell's Reports
Ch D	Law Reports, Chancery Division
CILL	Construction Industry Law Letter
CLD	Construction Law Digest
Const LJ	Construction Law Journal
Con LR	Construction Law Reports
C & P	Carrington & Payne's Reports
DLR	Dominion Law Reports
EG	Estates Gazette
Exch	Exchequer Reports
Ex D	Law Reports, Exchequer Division
Giff	Giffard's Reports
HBC	Hudson's Building Contracts
HLC	House of Lords Cases
KB	Law Reports, King's Bench Division
LGR	Local Government Reports
LJEx	Law Journal, Exchequer
LJQB	Law Journal, Queen's Bench
Lloyd's Rep	Lloyd's List Law Reports
LR/CP	Law Reports, Common Pleas
LT	Law Times Reports
M & W	Meeson & Welsby's Reports
NZLR	New Zealand Law Reports
QB	Queen's Bench Reports
SALR	South African Law Reports
TLR	Times Law Reports
TR	Term Reports
WLR	Weekly Law Reports
WR	Weekly Reporter

Table of clause references: ICE Conditions Sixth edition

Note: references are to chapter sections.

Clause:

Index

Note: references are to chapter sections.